THE RELATIONSHIP BETWEEN KNOWLEDGE TRANSFER, TEAM LEARNING, AND PROJECT SUCCESS IN THE INFORMATION TECHNOLOGY FIELD

by

Dixie D. O'Connell Overton, Ph.D., PMP

RAJ SINGH, PhD, Faculty Mentor and Chair
HUNG KIEU, PhD, Committee Member
JOSÉ NIEVES, PhD, Committee Member

Rhonda Capron, EdD, Dean of Technology
School of Business and Technology

A Dissertation Presented in Partial Fulfillment
Of the Requirements for the Degree
Doctor of Philosophy

Capella University
March 2017

Copyright © 2018 by Dixie D. O'Connell Overton, Ph.D., PMP

Library of Congress Control Number:		2018901628
ISBN:	Hardcover	978-1-5434-8354-3
	Softcover	978-1-5434-8353-6
	eBook	978-1-5434-8352-9

All rights reserved. No part of this book may be reproduced or transmitted in any form or by any means, electronic or mechanical, including photocopying, recording, or by any information storage and retrieval system, without permission in writing from the copyright owner.

Any people depicted in stock imagery provided by Getty Images are models, and such images are being used for illustrative purposes only.
Certain stock imagery © Getty Images.

Print information available on the last page.

Rev. date: 02/24/2018

Connect to the Author
Dixie@DrDixieOverton.com
Business Website: http://www.DrDixieOverton.com
LinkedIn Profile: https://www.linkedin.com/in/dixieoverton

To order additional copies of this book, contact:
Xlibris
1-888-795-4274
www.Xlibris.com
Orders@Xlibris.com
774421

Dedication

There is nothing more appropriate than dedicating this academic milestone to the late Orville and Joyce O'Connell. No matter what you study, or how long you go to school, you never quit learning from your parents. Mom and Dad were with me when I started this journey but are only with me in spirit and memories at the completion. I feel a great sense of comfort knowing they are looking down on me from heaven with extreme pride. Thank you for molding me into the person I am today. I love you. I miss you.

Dixie O'Connell Overton

Dixie O'Connell Overton

Table of Contents

List of Tables viii

List of Figures x

Abstract xi

Acknowledgments xiii

CHAPTER 1. INTRODUCTION 1

    Introduction to the Problem 1

    Background of the Problem 5

    Statement of the Problem 14

    Purpose of the Study 15

    Significance of the Study 16

    Research Questions and Hypotheses 17

    Definition of Terms 19

    Research Design 26

    Assumptions and Limitations 27

    Organization for Remainder of Study 29

CHAPTER 2. LITERATURE REVIEW 30

    Introduction 30

    Methods of Searching 30

    Theoretical Orientation for the Study 32

    Review of the Literature 47

    Findings 76

    Critique of Previous Research Methods 79

    Summary 82

CHAPTER 3. METHODOLOGY 84
    Introduction..84
    Purpose of the Study 84
    Research Questions and Hypotheses.......................................86
    Research Design...89
    Target Population and Sample...94
    Procedures...99
    Instrument..104
    Ethical Considerations 108
    Summary... 111
CHAPTER 4. RESULTS 112
    Background 112
    Description of the Sample 112
    Data Analysis and Results... 117
    Hypothesis Testing 124
    Summary..140
CHAPTER 5. DISCUSSION, IMPLICATIONS,
                RECOMMENDATIONS.................................... 141
    Introduction... 141
    Summary of the Results.. 141
    Discussion of the Results 144
    Conclusions Based on the Results...148
    Interpretation of the Findings..152
    Limitations 154

    Implications for Practice 155
    Recommendations for Further Research.............................. 158
    Conclusions 160
REFERENCES 162
STATEMENT OF ORIGINAL WORK... 178
INDEX 181

List of Tables

Table 1	Citations and Sources	46
Table 2	Demographic Characteristics	96
Table 3	Types of Data	106
Table 4	Age	113
Table 5	Project Roles	113
Table 6	Industry/Field in Which Project Was Conducted	114
Table 7	Size of Project Team	115
Table 8	Level of Project Complexity	115
Table 9	Descriptive Statistics	116
Table 10	Kolmogorov-Smirnov Test of Normality	117
Table 11	Reliability Coefficients	118
Table 12	Factor Score Weight Matrix	119
Table 13	Significance of Factor Structure	122
Table 14	Correlation Matrix	123
Table 15	Regression Weights for Research Question 1/ Hypothesis 1	127
Table 16	Regression Weights for Research Question 2/ Hypothesis 2	128
Table 17	Regression Weights for Research Question 3/ Hypothesis 3	129
Table 18	Regression Weights for Research Question 4/ Hypothesis 4	131
Table 19	Regression Weights for Research Question 5/ Hypothesis 5	133

Table 20	Regression Weights for Research Question 6/ Hypothesis 6	134
Table 21	Regression Weights for Research Question 7/ Hypothesis 7	136
Table 22	Regression Weights for Research Question 8/ Hypothesis 8	138
Table 23	Summary of Hypotheses Tested and Outcomes	139
Table 24	Comparison of Cronbach's Alpha	150

List of Figures

Figure 1. Structural model 91

Figure 2. Confirmatory factor analysis 120

Figure 3. Path diagram for Research Question 1/Hypothesis 1 ... 127

Figure 4. Path diagram for Research Question 2/Hypothesis 2 ... 128

Figure 5. Path diagram for Research Question 3/Hypothesis 3 ... 130

Figure 6. Path diagram for Research Question 4/Hypothesis 4 ... 132

Figure 7. Path diagram for Research Question 5/Hypothesis 5 133

Figure 8. Path diagram for Research Question 6/Hypothesis 6 ... 135

Figure 9. Path diagram for Research Question 7/Hypothesis 7 137

Figure 10. Path diagram for Research Question 8/Hypothesis 8 .. 138

Abstract

A quantitative survey design was used to understand the relationship of the dynamic capabilities of knowledge sharing and team learning to project success controlling for information technology projects in the United States. The literature review considered the theories of organizational learning and knowledge management to understand the development of dynamic capabilities in project management. A high project failure rate has been reported for decades, and the relationship between dynamic capabilities and project success has not been well researched. The overall research question was: *How does the model of knowledge transfer, team learning, and project success explain the relationship between project success and individual knowledge, knowledge articulation, and knowledge codification, controlling for the effects of information technology projects?* Structural equation modeling was used to test the model and address the research questions and hypothesis. The population consisted of project professionals across industries that had worked on an information technology project in the United States in the past year. The final sample included 128 fully completed surveys. Forty-six percent of the variance in project success was explained by the final model. Knowledge articulation had the lowest correlation to cross-project learning, indicating that individual knowledge and knowledge codification are more important for cross-project learning. The data supported the model indicating that knowledge transfer activities impact both project learning and cross-project learning which contributes to project success in the IT field. Recommendations for further research include controlling for the level of project management maturity, the extent of IT infrastructure, and whether the project team members were co-located.

Acknowledgments

There are numerous people who helped me in varying ways during this journey. First, I would like to thank my mentor, Dr. Raj Singh, whose guidance, support, understanding, and encouragement was exceptional. I would also like to thank my committee members, Dr. Hung Kieu and Dr. José Nieves; their suggestions and feedback were instrumental. The Capella Dissertation Writers Retreat was extremely helpful, and I appreciate the coaching and recommendations from Dr. Rubye Braye, Dr. Edward Mason, Martha Ruddy, and Dr. Richard Shrek.

I would also like to thank Dr. Linda Edelman and Dr. Sue Newell for allowing me to use their survey instrument, without which I would not have been able to complete this research. Special thanks to Dr. Julie Conzelmann for her editing and formatting expertise. In addition, another special thank you to Dr. Harold Whitfield for his statistical prowess, which was of vital assistance in conducting SEM analysis.

I have a great deal of love and appreciation for my friend and sister, Deb Watson Lentsch, who always believed in me and was understanding when I said I didn't have time to do fun things. Finally, I am grateful for my husband, Scott Whitney, for picking up the slack in the household chores and for sticking with me throughout this time. We were married only four months before I started this journey and are looking forward to a relationship that allows a little more time for each other.

CHAPTER 1. INTRODUCTION

Introduction to the Problem

There is a low success rate for information technology (IT) projects. The Standish Group has published the CHAOS Report every year since 1994. This report provides a snapshot of the status of IT projects in the software development realm. The 2015 report showed that the overall success rate had remained low, ranging from 27% to 31% for the last five years (Hastie & Wojewoda, 2015). In the same time period, the rate of challenged projects has varied from 49% to 56%. There has been very little change in project results, despite improvements in technology and in project management.

In addition, the failure rate of IT projects have seen limited improvements, varying only from 17% to 22% in the last 5 years (Hastie & Wojewoda, 2015; Levin, 2010). These ranges are also in agreement with the CHAOS Report published in 2001 (Lierni & Ribière, 2008), showing that no significant improvement was made in the past 15 years. Cerpa and Verner (2009) reported that projects fail for the same reasons they did 30 years ago. However, ways to increase project success have been suggested.

Effective knowledge management (KM) was shown to influence project success. Three of the top 10 factors identified by Hastie and Wojewoda (2015) that made IT projects more successful include individual knowledge, knowledge sharing, and knowledge transfer. Lack of individual and team knowledge is a risk to project performance (Lierni & Ribière, 2008); knowledge needs to be shared and successfully integrated (Levin, 2010). Gasik (2011) agreed that project knowledge management (KM) is one of the main success factors in project management. Knowledge management is particularly important for IT projects.

Greater challenges exist when the project involves a high degree of technology or is dissimilar to past projects, which is common in the IT field. Without knowledge transfer and sharing, organizations fail to continue practices that worked well, and fail to discontinue those that resulted in errors or rework, as evidenced in the findings of The Standish Group. In addition, KM is an important component of project team learning, which was identified by Jetu and Riedl (2012) as one of three main focuses that influence project success in IT projects. The goal of this study is an attempt to understand the importance of, and how, knowledge sharing can contribute to IT project success rates.

The project management field is increasing and is expanding to more industries. Between 2010 and 2020, the demand for project management professions is expected to grow 12% in the United States, according to the Project Management Gap Report (Project Management Institute [PMI], 2013). The report identified business services, manufacturing, finance and insurance, oil and gas, information services, construction, health care, and utilities as the leading project intensive industries. Projects are prioritized and initiated to meet an organization's strategic goals (Levin, 2010). Therefore, learning how to improve success rates may benefit organizations and practitioners in many industry sectors.

"All change in an organization happens through projects and programs" (Project Management Institute [PMI], 2016, p. 14). The need for experienced and effective project managers can only increase as more and more changes are planned for organizations and industries. Of great importance to both scholars and practitioners is the need to identify skills and practices that increase IT project success rates.

The number of organizations that have a defined training plan and career path for project managers has remained the same since 2012 (PMI, 2016). Moreover, less than half of organizations surveyed by Project Management Institute (PMI) have a formal knowledge transfer process, which has actually decreased 5% from 2015 to 2016. Consequently, more projects are failing, resulting in monetary loss for the organization. Specifically, the report disclosed that $122 million is wasted for every $1 billion invested because of poor project performance; an increase of 12% from 2015 to 2016 (PMI, 2016).

In the early 21st century, research in project management expanded to include concepts such as organizational learning and knowledge management (Morris, 2013). One of the main success factors in project management, project knowledge management, was first mentioned in the literature in 1987 (Gasik, 2011). However, knowledge management has not received the attention as other areas in project management, such as risk management, quality management, or communications management (Gasik, 2011). The *Project Management Book of Knowledge* (PMBOK), the leading guidebook for project management practices in the United States, lists 47 explicit knowledge objects, emphasizing the importance of knowledge in project management (Reich & Wee, 2006).

While there is no specific definition of a project team (Jetu & Riedl, 2012), it is characterized by members that are discontinuous and may have separate goals. Therefore, project team learning is challenged because of the temporary and time constrained nature of projects. Levin (2010) also pointed out that there is no standard definition of knowledge management, which makes communicating its importance difficult. Gasik (2011) listed seven definitions of project knowledge management in two separate categories. Few studies link learning within the project team and learning across project teams, or from the project to the larger organization (Swan, Scarbrough, & Newell, 2010).

The results of this study should help project professionals and project stakeholders on defining beneficial methods for knowledge sharing to help increase the success rates, and lower the costs of IT projects. The expectation is that this study will also contribute to the field of project management in the knowledge management realm. Knowledge management is not well defined and there is a gap in the research concerning knowledge management and practical project management practices. The information from this research will also help identify additional research that can contribute specifically to knowledge management and the project manager profession.

The remaining sections of this initial chapter will describe the background of the study, the statement of the problem, and the purpose and significance of this study. The research questions will be presented, followed by definitions of terms used in this research. An overview of the research design will follow, with more detail of the design presented in Chapter 3. Assumptions and limitations will be listed near the end of this chapter.

Background of the Problem

Project management is a young discipline. It was not established in the literature in the form known today until the mid-1980s. There is no overarching theory of project management; however, it is an emperical discipline that occurs within and between organizations (Jugdev & Mathur, 2013). Therefore organizational learning theories are appropriate to provide theoretical guidance to the project management field (Jugdev & Mathur, 2013).

The theoretical domain of organizational learning includes knowledge management (Jugdev, 2012). Seminal works in organizational learning emerged in 1965, but knowledge management research did not receive much focus until the 1990s (Reich, 2007). Steyn and Kahn (2008) stated that while organizations are increasing aware of knowledge management to increase competitive advantage, there is still no predominant knowledge management theory. Steyn and Kahn (2008) suggested this may be because of knowledge management spanning many disciplines, such as learning, knowledge practices, knowledge processes, and knowledge manipulation.

Knowledge management is an appropriate concept in the project management disclipline. "Knowledge management in the context of a project is the application of principles and processes designed to make relevant knowledge available to the project team" (Reich, 2007, p.8). Jackson and Klobas (2008) claimed that projects "generate the personal and group knowledge which contribute to their own success" (p. 329). Knowledge management includes the who, what, when, and how of information flow (Lierni & Ribière, 2008; Singh & Soltani, 2010).

Organizational learning is also appropriate for the IT field. Organizational learning is influenced by stresses commonly occurring in IT project management. These stresses include the complexity and uncertainty of the environment, and the stress resulting from the effort trying to understand it, and to predict the future (Cangelosi & Dill, 1965). This stress is common in IT project teams, especially if they are working on new technologies, and struggling to achieve project success, when many projects still fail.

The creation and consumption of knowledge manifest in changes of practice for the individual and for the collective. The theory of organizational learning addresses learning at the individual, team, and organizational level, as reported in the seminal work by Cangelosi and Dill in 1965 (Crossan, Lane, & White, 1999; Haldeman, 2011; Reich, 2007). Organizational learning theory supports the construct of team learning.

Organizational learning is context-based; it depends on the actions of individuals and situations in which they occur. Senge (1994) defined a learning organization as a community that promotes continual learning. Situational learning theory was introduced by Lave and Wenger (1991). Situational learning theory expands on organizational learning from an action perspective, emphasizing that learning occurs based on actions and practices individuals develop to fit within the community and have a sense of identity (von Krogh, 2002). Situational learning is informal learning embedded in practice, context, and culture (Jugdev & Mathur, 2013). Situational learning theory and behavior theory have similarities to agency theory.

Agency theory focuses on the interactions of agents and principals. Agency theory was applied in managment but not extensively in IT (Mahaney & Lederer, 2011). Mahaney and Lederer (2011) surveyed developers to determine the interactions between privately held information, project monitoring, and project success. The findings supported that project monitoring practices increase knowledge transfer, and positively influence project success.

The sensitivity to success and failure felt inside the organization was refered to by Cangelosi and Dill (1965) as performance stress. Performance stress is another influential factor to organizational learning. The pressures of project success increase when the project is complex, or has technical challenges (Jugdev & Mathur, 2013). Performance stress is expected to be high in IT project management. Therefore, organizational learning theory supports the use of the construct of project success.

Organizational learning and knowledge management both address team learning. Knowledge management can be illustrated by learning within and across levels of the organization (Crossan, Maurer, & White, 2011). Jugdev and Mathur (2013) recognized a gap in the literature regarding applying the concepts of organizational learning in the project management field, and that organizational competitiveness can be improved through effective project learning and cross-project learning. Project team learning and cross-project team learning are the compontents of the team learning construct in this study.

Team learning is influenced by individual knowledge and experience (Cangelosi & Dill, 1965). For individual knowledge to be amplified it needs to be articulated, therefore it is critical for organizations to have a process in which team members can interact (Nonaka, 1994). Nonaka (1994) establised an organizational knowledge creation theory and framework and found that new knowledge is developed by individuals and is based on their experiences and what they believe to be true. The interactions of individuals is necessary, not only for creating knowledge, but for transferring knowledge as well.

Some factors of the theory of knowledge transfer involve a knowledge sharing process, incentives, and process facilitation (Markus, 2001). Markus (2001) created an initial step toward developing a theory of knowledge transfer, referred to in the article as knowledge reuse. Knowledge transfer should positively influence project success but there are challenges in a project environment.

One of the challenges of knowledge transfer in projects may occur because of the belief that the project is unique and that knowledge transfer is not needed. The theory of planned behavior explains that actions and intentions are based on perceived outcome (Jewels & Ford, 2006; Stewart, May, & Ledgerwood, 2015). Newell and Edelman (2008) confirmed that documenting learning, for example, is not perceived as beneficial. The behavior approach to organizational learning acknowledges that individuals and groups often fail to understand how deliberate knowledge sharing can influence project and organizational outcomes (Cangelosi & Dill, 1965).

Paramkusham and Gordon (2013) agreed that knowledge transfer could improve project performance. Project performance depends on the level of effort of knowledge transfer (Landaeta, 2008). Paramkusham and Gordon (2013) suggested future research in the area of knowledge transfer and project performance.

Factors of knowledge transfer and project performance can be researched based on dynamic capabilities. Dynamic capabilities are based on the theory of organizational learning and are an expansion of the concepts in the knowledge creation theory (Teece, Pisano, & Shuen, 1997; Zollo & Winter, 2002). Zollo and Winter (2002) defined a dynamic capability as a "learned and stable pattern of collective activitiy through which the organization systematically generates and modifies its operating routines in pursuit of improved effectiveness" (p. 340).

While the seminal paper on dynamic capabilities considers them an ability (Teece et al., 1997), they can also be considered a process (Zollo & Winter, 2002), which is easier to measure (Di Stefano, Peteraf, & Verona, 2014). Zollo and Winter (2002) identified three learning mechanisms, individual knowledge, knowledge articulation, and knowledge codification, to understand how deliberate cognitive activity shapes organizational learning. Individual knowledge, knowledge articulation, and knowledge codification are the three variables that make up knowledge transfer in this research.

Newell and Edelman (2008) considered the three learning mechanisms when surveying a utilitiy company in the United Kingdom. Newell and Edelman (2008) acknowledged that there are few emperical studies focused on the development of dynamic capabilities and organizational learning. This research will expand on this concept as it is related specifically to project management in the IT field.

A motivation for undertaking this research is direct experience with some IT projects that have gone well, some that have been challenged, and some that failed; thus, the projects were restructured for various reasons. With regard to project management, the goal for conducting this study was to discover the role of knowledge transfer that may be controlled by the project manager to influence project success. Both practitioners and scholars are interested in this research because of the growing field of project management.

Another expectation for conducting this study was to determine if knowledge transfer and team learning contribute to the success of information technology projects. Variables in this study, such as team learning, and knowledge transfer, fall under social research as they are variables dealing with people and cannot be directly measured. Rovai, Baker, and Ponton (2014) defined social research as research focused on people to increase the understanding of human behavior. Survey research design is a common non-experimental method for social research.

Surveys composed of questions that can be grouped together to represent each variable to be measured can be used to conduct quantitative statistical analysis. During the course of this quantitative study, a research design of surveying will help gather data to answer the research questions. Likert-type responses can be converted into integers to obtain interval data (Vogt, 2007). The research questions and hypothesis in this survey ask about the relationship, if any, between variables. Regression analysis will be the statistical method to analyze the direction of the relationship between variables (Cooper & Schnider, 2011). Confirmatory factor analysis will also determine if the survey questions have been condensed and are reliable measures of the variables (Sproull, 2003).

In a non-experimental study it is not possible to conclude cause and effect since there is no control placed in the natural setting of the project execution (Sproull, 2003). The conclusions from this study will help to measure associations. The approach for this study is predictive and will make an attempt to predict how the independent variables relate to the dependent variables (Cooper & Schnider, 2011).

Information technology (IT) projects implement change to bring value to the organization. Value to the organization in implementing projects is primarily knowledge work; combining technology knowledge and business process knowledge to achieve project goals (Reich, Gemino, & Sauer, 2008). Knowledge transfer is beneficial for the high degree of complexity and integration typical of IT projects (Crawford & Pollack, 2007; Gemino, Reich, & Sauer, 2008). However, knowledge transfer and learning can be challenging in the project environment.

The temporary and discontinuous structure of projects provides a challenge to learning (Landaeta, 2008). A lack of learning from past or current projects can lead to repeating the same mistakes (Cerpa & Verner, 2009; Jugdev, 2012; Newell & Edelman, 2008). This lack of learning is partly because of believing that the project is too unique to supply knowledge that could be used for a different project (Newell

& Edelman, 2008). Increasingly effective knowledge management practices are needed in the project management field.

Knowledge management is a key component of project management and organizational learning (Crawford & Pollack, 2007; Goffin & Koners, 2011; Lierni & Ribière, 2008; Singh & Soltani, 2010). Effective knowledge management in IT projects should increase project success rates (Cerpa & Verner, 2009; Jewels & Ford, 2006). There are many articles on conceptual knowledge management and organizational learning, but there is a gap in the literature when it comes to practical project management (Reich, 2007). One method of knowledge management in project management is documentation.

Project team members participate in knowledge transfer by documenting lessons learned or conducting post project reviews. These reviews may not be done because of the belief that they will not be helpful (Newell & Edelman, 2008). Organizational learning and project performance are hindered by a lack of lessons learned; this was initially reported in the literature almost 50 years ago, and is still true today (Reich, 2007). Jugdev (2012) acknowledged that the topic of lessons learned, the learning gained during the project, is emerging in the literature and is under represented in conferences. The process for conducting this study includes knowledge documentation, referred to as knowledge codification, as one of the constructs of knowledge transfer.

Statement of the Problem

Information technology projects are implemented in organizations to achieve strategic goals. The problem is the failure rate of IT projects continues to remain high, costing companies billions of dollars and missed opportunities. Information technology projects are mainly knowledge work, where the project brings together knowledge of technology and business process to create value to the organization (Gemino et al., 2008). Additional research is needed to explore the organizational learning and knowledge management impacts to organizational goals and capability development through the project management field.

There is a gap in the literature relating project team learning to the overall organization (Swan et al., 2010). Additionally, Jugdev and Mathur (2013) identified a gap in the literature relating organizational learning in the project management field. Further research was suggested by Paramkusham and Gordon (2013) in the area of knowledge transfer and project performance.

Newell and Edelman (2008) recommended additional research regarding organizational learning and the development of dynamic capabilities. The goal of this research was to use the three components of dynamic capabilities framework: individual knowledge, knowledge articulation, and knowledge codification, and identify how they impact project team learning in ways that may help improve IT project success rates.

Purpose of the Study

The purpose of this regression survey research was to test the model of knowledge transfer, team learning, and project success that relates individual knowledge, knowledge articulation, and knowledge codification to project success controlling for projects in the IT field. Knowledge is the most important project management resource, and knowledge management is required for effectively managing projects (Gasik, 2011). Effective project learning can reduce repetitive mistakes and contribute to business success (Jugdev & Mathur, 2013). It is important to understand how to increase project effectiveness, especially for emerging industries that may not have the background in IT.

Project management is a growing field and IT projects span a diverse number of industries (PMI, 2013; The Standish Group, 2013). The results from conducting this study will be important to project management practitioners in all industries because improving project success rates will help reduce wasted money and missed opportunities. Conducting this study is relevant and timely to recent practitioner practices because of many limitations pointed out in the PMBOK, the leading guide for project management practices (Abyad, 2012; Gasik, 2011; Koskela & Howell, 2008; Morris, 2013)

The findings from this study may also be important to scholars. In recent years, peer- reviewed articles in scholarly journals steadily increased awareness in the area of knowledge management, with an average of 261 articles per year between 2010 and 2015; 245 articles per year between 2006 and 2010; and 125 per year between 2001 and 2005. The concept of dynamic capabilities is also growing in popularity in research (Di Stefano et al., 2014). The increased trend and number of published articles shows a strong interest in this area. Without the information gained from conducting this study, the opportunity is missed to understand more about knowledge management as it relates to project management in the IT field across industries.

Significance of the Study

The expectation was that the findings from this study would add value to project management professionals, particularly for project managers working on IT projects in the United States, by providing more understanding of the effectiveness and efficiency of knowledge transfer methods that may contribute to project success. Project managers implementing knowledge transfer methods that contribute to project success will add value to project stakeholders and organizations that implement IT projects to meet strategic goals. The knowledge transfer methods identified would contribute to scholars by identifying areas of study that need further attention in project learning environments, specifically in the IT field.

The significance of this study was its contribution to the field of project management and knowledge management by exposing project management and knowledge management practices that may improve

IT project performance and organizational learning. The results from this study will benefit all practitioners that influence project performance in IT. The findings from this study will benefit project managers, project stakeholders, and organizations that have a high degree of IT projects by identifying practices that may contribute to effective knowledge management.

Research Questions and Hypotheses

How does the model of knowledge transfer, team learning, and project success explain the relationship between project success and individual knowledge, knowledge articulation, and knowledge codification, controlling for the effects of information technology projects? A regression analysis was used to test the overall hypothesis and determine the degree of the relationship between the variables.

RES Q1: What is the relationship, if any, between individual knowledge and project learning?

$H1_0$: There is no significant relationship between individual knowledge and project learning.

$H1_A$: There is a significant relationship between individual knowledge and project learning.

RES Q2: What is the relationship, if any, between individual knowledge and cross-project learning?

$H2_0$: There is no significant relationship between individual knowledge and cross-project learning.

$H2_A$: There is a significant relationship between individual knowledge and cross-project learning.

RES Q3: What is the relationship, if any, between knowledge articulation and project learning?

$H3_0$: There is no significant relationship between knowledge articulation and project learning.

$H3_A$: There is a significant relationship between knowledge articulation and project learning.

RES Q4: What is the relationship, if any, between knowledge articulation and cross-project learning?

$H4_0$: There is no significant relationship between knowledge articulation and cross-project learning.

$H4_A$: There is a significant relationship between knowledge articulation and cross-project learning.

RES Q5: What is the relationship, if any, between knowledge codification and project learning?

$H5_0$: There is no significant relationship between knowledge codification and project learning.

$H5_A$: There is a significant relationship between knowledge codification and project learning.

RES Q6: What is the relationship, if any, between knowledge codification and cross-project learning?

$H6_0$: There is no significant relationship between knowledge codification and cross-project learning.

$H6_A$: There is a significant relationship between knowledge codification and cross-project learning.

RES Q7: What is the relationship, if any, between project learning and project success?

$H7_0$: There is no significant relationship between project learning and project success.

$H7_A$: There is a significant relationship between project learning and project success.

RES Q8: What is the relationship, if any, between cross-project learning and project success?

$H8_0$: There is no significant relationship between cross-project learning and project success.

$H8_A$: There is a significant relationship between cross-project learning and project success.

Definition of Terms

Age. Age is a demographic variable collected in this study. Age ranges were identified as between 18-29 years, 30-44 years, 45-59 years, and 60 years or greater.

Cross-project learning. Cross-project learning is a mediating variable in this study. "Cross-project learning examines perceptions of the movement of learning across team boundaries to other project teams" (Newell & Edelman, 2008, p. 575).

Dynamic capabilities. "A dynamic capability is a learned and stable pattern of collective activity through which the organization systematically generates and modifies its operating routines in pursuit of improved effectiveness" (Zollo & Winter, 2002, p. 340).

Experience. The number of years the survey participant had been working on IT projects was collected as a demographic variable to measure the individual's experience. This was an open field where respondents could enter the number of years they have been working on IT projects. This data was collected as a demographic variable to compare the sample with the population demographics collected in the annual salary survey conducted by PMI (Project Management Institute [PMI], 2015).

Experience Accumulation. (See Individual knowledge.)

Explicit knowledge. Explicit knowledge takes the form of routines and techniques that can be formally codified in training manuals, lessons learned documents, etc.; often referred to as "know what" knowledge (Luhman & Cunlifffe, 2013, p. 127).

Gender. Gender is a demographic variable collected in this study limited to two values, male or female.

Individual knowledge. Individual knowledge is an independent variable in this study, referred to as experience accumulation in other studies (Newell & Edelman, 2008; Zollo & Winter, 2002). Based on the research by Senge (1994), Newell and Edelman (2008) explained that individual knowledge "examines the extent to which respondents perceived that learning is shared through the movement of people across projects" (p. 574).

Industry. The industry the project was conducted in was a demographic variable collected during this study. Selections provided were health care, non-profit, technology-information services, energy and utilities, transportation, construction, finance and insurance, government, professional services, manufacturing, education, or other. These selections were recommended by Qualtrics. Survey respondents were asked to specify their industry if they selected other. Project failure rates are similar across industries (Rungi, 2014), this research did not focus on a particular industry but rather on the IT field. Industry was collected as a demographic variable to compare the sample with the population demographics collected in the annual salary survey conducted by PMI (PMI, 2015).

Information. Information is a flow of messages with content that might add to, restructure, or change knowledge (Nonaka, 1994). Holsapple (2003) defined information as data that were organized for a specific use, such as process, sales, inventories, etc.

Information technology (IT). Information technology involves technology components, such as software, hardware, and infrastructure. Business processes, management, developers, users, and activities involving technology components are also considered part of IT (Steenkamp & McCord, 2003). Information technology is comprised of IT infrastructure, the organization's financial commitment to IT technical and human resources; IT-business relationships, the interactions between the business and the IT resources; and IT-business knowledge, how well the organization can apply IT to meet strategic goals (Crawford, Leonard, & Jones, 2011).

Information technology project. An information technology project is a project that has a combination of technical and human resources "to accomplish organizational objectives through the structuring of people, technology and knowledge content" (Davenport, De Long, & Beers, 1998, p. 44).

Knowledge. Knowledge is a justified true belief held by individuals, created by the flow of information (Nonaka, 1994). Knowledge is a blend of information, experience, ideas, and insights that guide behavior, actions, and judgment decisions; information becomes knowledge when it is understood and utilized by individuals (Luhman & Cunlifffe, 2013). Knowledge is based on information, however, gaining knowledge from information requires judgment, and is based on the context (situation) and individual experience (Holsapple, 2003).

Knowledge articulation. Knowledge articulation is an independent variable in this study. Based on the research by Zollo and Winter (2002), Newell and Edelman (2008) offered the following definition. "Knowledge articulation examines the extent to which respondents perceived that learning is captured and shared by deliberate project meeting and review processes, in which the team figures out what works and what does not in the execution of an organizational task" (p. 574).

Knowledge codification. Knowledge codification is "the inscription of knowledge into text, drawings, templates, models and similar media, often playing a central role in the strategies devised by firms to preserve and transfer learning" (Cacciatori, Tamoschus, & Grabher, 2011, p. 311). Knowledge codification is an independent variable in this study and is often referred to as knowledge documentation. Based on the research by Zollo and Winter (2002), Newell and Edelman (2008) offered the following definition "knowledge codification examines the extent to which respondents perceived that learning was captured and shared by documenting lessons" (p. 574).

Knowledge documentation. (See knowledge codification).

Knowledge management (KM). Knowledge management is the process of actively managing and leveraging knowledge of individuals and project teams, gained through skills and experience, to enhance organizational learning and project performance. "KM consists of a set of techniques and tools to make the right knowledge available to the right people at the right moment" (Scarso & Bolisani, 2011, p. 62).

Knowledge sharing. Knowledge sharing is the process of identifying knowledge creation that was validated by its application, and documenting or codifying that knowledge so it can be used by someone other than its creator (Gasik, 2011).

Knowledge transfer. Knowledge transfer is the act of communicating knowledge between a sender and a receiver. The sender and receiver can be individuals or teams. The method of communication can be codified or non-codified (Gasik, 2011). Markus (2001) used the phrase knowledge reuse as synonymous with knowledge transfer. In this study, knowledge transfer is the construct comprising of the three independent variables, individual knowledge, knowledge articulation, and knowledge codification.

Learning. Definitions of learning were often lacking in the articles from 1999-2004 reviewed by Fenwick (2008). The term learning may refer to a process as well as an outcome (Fenwick, 2008).

Lessons learned. Lessons learned represent "the learning gained from the process of performing the project" (Project Management Institute [PMI], 2008, p. 436)

Level of project complexity. The level of complexity was a demographic variable included in this study. Five measures of complexity were offered: simple-repeatable change, very similar to other projects, somewhat similar to other projects, slightly similar to other projects, extremely complex-not done before.

Organizational learning. Organizational learning is realized by changing organizational actions through reflection on new knowledge and understanding gained by subgroups and teams (Swan et al., 2010).

Product versus Service. A demographic variable collected in this study includes whether the project implemented a project or a service.

Project. According to Project Management Institute's (2008) A Guide to the Project Management® Body of Knowledge, a project is "a temporary endeavor undertaken to create a unique product, service, or result" (p. 434).

Project learning. Project learning was referred to as team learning by Newell and Edelman (2008) and defined as "...the extent to which the team has changed the way it operates based on knowledge gained" (p. 574). Project learning is a mediating variable in this study.

Project management. Project management is "the application of knowledge, skills, tools, and techniques to project activities to meet the project requirements" (PMI, 2008, p. 435).

Project roles. Project roles was a demographic variable collected in this study. Participants were asked to select which option best described their project role. Options included business project manager, technical project manager, business project team, technology project team, business stakeholder, technology stakeholder, project sponsor, executive project sponsor, or other. Survey respondents were asked to specify their project role if they selected other.

Project success. Project success is measured by the degree the project delivered on time, on budget, and with required features and functions (The Standish Group, 2013). In this study, project success is the independent variable.

Project team. A project team is a group of individuals that are temporarily horizontally aligned (function-oriented), rather than vertically aligned (top-down-oriented), to achieve a goal. Team members are heterogeneous on their knowledge, as individuals are selected for the project based on their skills and experience (Gonzalez & Martins, 2014).

Size of project team. The size of the project team was a demographic variable collected during this study, representing the number of individuals assigned to the project. There were six size ranges to select from: 1, 11-25, 26-50, 50-100, 100-200, and over 200.

Tacit knowledge. Tacit knowledge takes the form of intuition and improvisation that is difficult to codify because of its personal and context-dependent nature, but may be articulated through stories and metaphors; often referred to as "know how" knowledge (Polanyi, 1966; Teece, 2007, p. 1339). Riding a bicycle is a common example of tacit knowledge.

Team learning. Team learning in this study is a construct comprising the mediating variables of project learning and cross-project learning.

Research Design

This non-experimental research used an online survey approach to obtain quantitative data. Newell and Edelman (2008) developed the survey used in this study. Newell and Edelman (2008) studied the dynamic capabilities framework established by Zollo and Winter (2002) and sought to understand how organizations and projects benefit from project learning. Newell and Edelman (2008) focused on a utility company in the United Kingdom; this research expands the study to IT projects in the United States. The IT field is growing and challenged because of the complexity and uniqueness of IT projects.

The survey questions were entered in Qualtrics, and the survey was distributed to a random sample of their audience. The qualifying question was whether participants had worked on an IT project in the

United States within the past year. The survey was sent out once and 128 complete responses were obtained.

Structural equation modeling (SEM) was used to analyze the data. Structural equation modeling is a form of regression analysis and is a common method of quantitative research when testing a model. Structural equation modeling has been applied by other researchers in the project management field (Lichtenthaler, 2009; Mahaney & Lederer, 2011; Newell & Edelman, 2008; Zheng, Zhang, & Du, 2011).

Assumptions and Limitations

Assumptions

Assumptions are expectations regarding the research that may increase risk and bias to the research if not controlled. Assumptions in this research related to the survey instrument, the survey respondents, third parties, and data analysis. In this study, it was assumed that the survey would adequately measure the variables addressed, and that the survey was worded in such a way that respondents clearly understood the survey questions and terms used. The use of an existing survey helped mitigate these risks. It was also assumed that the respondents had the knowledge and experience to provide useful and accurate responses, and that they were willing to do so.

Additional assumptions not under control of the researcher included the conduct of third parties. In this research, Qualtrics was used to distribute the survey and collect data. One assumption was that Qualtrics sent the survey out to the entire group and did not target specific areas. Another assumption was that Qualtrics will keep the data encrypted and stored in a manner consistent with their policies. Likewise, the data collected should not include confidential or identifying information. The researcher followed Capella University's IRB guidelines to control risks in these areas.

Assumptions regarding the data analysis included the sample size. It was assumed that the sample size was large enough to represent the population adequately and that the responses represented a good cross-section of IT project professionals. This allowed the results to apply to the general population of IT project professionals.

Limitations

Limitations represent risk or issues that prevent the research from being 100% accurate and reliable. Limitations, or constraints, in this study involved the survey respondents, third parties, and data analysis. It was a limitation that survey participants could base their responses on what actually happened during the project, and not what they thought should have happened, which would introduce bias. Respondents could have been reluctant to report on failed projects. However, Cerpa and Verner (2009) obtained results that included 70 failed projects. The reliance on third parties, such as Qualtrics, was a known constraint or limitation to this study.

Factors that could have influenced project performance not included in this research were overall organizational culture and whether the project team was co-located or virtual. The research survey instrument is regarded as one of the advanced instruments in dynamic capabilities research (Eriksson, 2013). The survey includes a section asking respondents to consider a project that was completed, thus it occurred in the past. This was warranted because project success could not be determined until the project was completed. Since dynamic capabilities develop over time, the focus of this research did not address the current state of the organization.

Organization for Remainder of Study

Chapter 2 of this manuscript consists of a review of the literature that served as a foundation for this study. Chapter 3 contains details regarding the research design, population, survey instrument, data collection, and data analysis method. The results are presented in Chapter 4. A discussion of the findings, with recommendations for future research, will be offered in Chapter 5.

CHAPTER 2. LITERATURE REVIEW

Introduction

The theoretical framework for this research of knowledge transfer and team learning in the context of IT projects is presented in this chapter. The review will begin with the work of seminal authors in the field of organizational learning. Developments in this field include organizational learning, knowledge management, and dynamic capabilities. Viewpoints in the literature regarding knowledge creation theory and situated learning will be compared and contrasted in the area of project management. Finally, recent work supporting the variable constructs used in this research is presented. The remainder of this chapter will include the library search methods to locate articles, and the seminal and evolution of the theoretical orientation for this study. A review of the literature is followed by a synthesis of the research findings as well as a critique of the previous research methods.

Methods of Searching

The majority of articles reviewed were found in the Business Source Complete database. All articles are from peer-reviewed journals. A few articles were located in the ABI/Inform Global database and Oxford Handbook Online. The Oxford Handbook Online provided useful concept papers and histories of theories after searching for project management and a separate search for organizational learning.

The article by Newell and Edelman (2008) that operationalized dynamic capabilities as defined by Zollo and Winter (2002) was familiar to the researcher at the start of this project; therefore, search phrases commonly used included the terms dynamic capabilities and project management. Other search phrases included variations of the names of the mediating variables (team learning or project team learning, and cross-project learning). The dependent variable names were also used individually or combined with project management (individual knowledge, knowledge articulation, and knowledge codification). To expand on the concept of dynamic capabilities, this phrase was used singly. The combination of project learning and IT projects provided several useful articles.

The resulting articles from these searches provided additional phrases that were used individually, such as knowledge creation theory and theory of knowledge management. Surprisingly some search combinations did not provide many useful articles, such as PMBOK and knowledge management and project management and knowledge management. Finally, Google Scholar was used to ascertain which articles were more influential based on the number of times they had been cited in the literature. A few books that were mentioned several times in the literature were included to provide basic concepts, seminal information, and definitions.

Theoretical Orientation for the Study

Introduction

Projects have been around since the beginning of time; yet, modern project management is a young discipline with no overarching theory. However, the foundational theory of organizational learning can be applied. Organizational learning theory includes the resource-based view (RBV) or theory of the firm, and knowledge management. Building on the resource-based view and knowledge management, organizational learning developed theories of knowledge, and the knowledge-based view of the firm. Dynamic capabilities extend the resource-based view, and represent a theory of strategic management concerning change management to gain competitive advantage. Since change occurs in organizations through the implementation of projects, it is appropriate to apply dynamic capabilities to project management. The strategy management theories of the resource-based view and dynamic capabilities are closely aligned with concepts of organizational project management (Drouin & Jugdev, 2013). Dynamic capabilities focus on the continuous rebuilding of and manipulating resources, which applies to project management because project management teams are not static. Project teams are aligned together on a temporary basis to achieve a collective goal. It is also appropriate to apply dynamic capabilities to the IT field because of the rapid advance and changes in technology that affect organizations creating a dynamic environment. Information technology projects deal with changing environments with dynamic business goals with increasing difficulty and competition (Paramkusham & Gordon, 2013).

Organizational Learning Theory

Organizational learning was first introduced in the literature by Cyert and March in 1963, and expanded by Cangelosi and Dill in 1965. However, it was not until the mid-1990s that it was developed further by important authors such as Peter Senge in the United States, and Ikujiro Nonaka and Hirotaka Takeuchi in Japan, among others (Luhman & Cunlifffe, 2013). However, no formal theory of organizational learning was developed (Crossan et al., 2011).

The objective of organizational learning is to create an environment within the organization that supports and encourages knowledge creation and knowledge sharing; learning opportunities that benefit the organization by increasing competitive advantage. This is not the same as normal training and development that is focused on individuals (Luhman & Cunlifffe, 2013). Senge defined learning organization as "organizations where people continually expand their capacity to create the results they truly desire, where new and expansive patterns of thinking are nurtured, where collective aspiration is set free, and where people are continually learning how to learn together" (Senge, 1994, p. 3). Senge's work was highly regarded and popular with both scholars and practitioners, partly because of his critique of bureaucratic corporations and top-down management (Caldwell, 2012).

Organizational learning was around for decades but the concept of knowledge was popularized in the mid-1990s. Further development came from a territorial debate. The initial focus on knowledge management concerned the technical aspects of IT; the development of databases and knowledge warehousing (Easterby-Smith, Crossan, & Nicolini, 2000). The technical aspects of IT converged with the social aspects of organizational learning and more emphasis was placed on social factors of knowledge management. Knowledge assets have been classified as hard, such as databases, knowledge repositories, and other technological knowledge management tools, and soft, such as culture, trust, values, and routines (von Krogh, Nonaka, & Rechsteiner, 2012). Even though knowledge management is a relatively new field of study, there are already over 100 published definitions spanning several domains (Scurtu & Neamtu, 2015).

Resource-Based View and Knowledge Management

The resource-based view (RBV), initially established by Penrose (1959), recognized the importance of an organization's core competencies, and stated that similar organizations can improve performance by how leaders manage strategic resources. The resource-based view stated that sustained competitive advantage can occur when resources are valuable, rare, inimitable, and non-substitutable (VRIN) (Eisenhardt & Martin, 2000; Lin & Wu, 2014; Peteraf, Di Stefano, & Verona, 2013; Teece, 2014; Wu, 2010).

Knowledge as a competency is essentially the resource-based view of the firm from a strategic point of view. The resource-based theory has evolved into the knowledge-based theory which considers resources as knowledge assets (Nonaka & Von Krogh, 2009). The knowledge-based view treats knowledge as a critical source of competitive advantage (Dinur, 2011). Knowledge must be shared to be useful, however, sharing increases the risk of imitation, and thus the risk of reduced competitive advantage. As the knowledge sharing increases, there is a greater need to simplify knowledge and make it more explicit (Newell, Bresnen, Edelman, Scarbrough, & Swan, 2006). This simplification results in a greater risk of imitation, resulting in a sharing versus protecting paradox (Dinur, 2011). However, knowledge alone does not imply capabilities.

Knowledge is considered the most important organizational resource of the 21st century that is able to bring long-term sustainable advantage to the organization (Gonzalez & Martins, 2014). Knowledge management considers knowledge as the most valuable resource that should be created and put into action (Tzortzaki & Mihiotis, 2014). The distinction between internal (tacit) knowledge, and external (explicit) knowledge, is from the seminal work by Polanyi in 1966, and is crucial to understanding concepts of knowledge management. Tacit knowledge is internal knowledge that we cannot explain or articulate. Explicit knowledge is external knowledge that can easily be taught or written down. Tacit knowledge is manifest in things we are attending from, in order to attend to explicit knowledge (Polanyi,

1966). People recognize a face because of the features of the face that we cannot readily describe (Polanyi, 1966).

A common example of tacit and explicit knowledge is reading a book about how to ride a bicycle, versus actually riding a bicycle. By reading the book, one gains explicit knowledge but not the tacit knowledge of how fast to pedal, and how to keep one's balance, which represent tacit knowledge. Dinur (2011) classified tacit knowledge into nine categories and found that documentation (codification) is only a useful knowledge transfer channel for four. More useful channels to transfer tacit knowledge include dialogue (articulation), hands on practice, observation, social interaction, and apprenticeship (Dinur, 2011).

Three Themes to Organizational Learning

There are three themes in the field of organizational learning and knowledge management: The SECI-Ba model, the social and practice perspectives, and dynamic capabilities (Luhman & Cunlifffe, 2013). The three themes focus on how knowledge is converted from tacit to explicit, and then internalized by the individual. Individual interactions and social practices are important for the environment in which learning takes place. The environment in which learning occurs can change, or be full of contradictions, that can enhance or focus the learning capability. The three themes will be presented next.

SECI-BA model. The SECI model represents a spiral of knowledge processes that can convert individual level knowledge into organizational level knowledge (Dinur, 2011). Its explanatory framework includes processes, knowledge assets, and organizational context (von Krogh et al., 2012). The SECI model was developed in Japan by Nonaka and Takeuchi in 1995, in their seminal work in the field of knowledge management. SECI represents four processes by which tacit and explicit knowledge are converted or combined: socialization, externalization, combination, and internalization (SECI).

Knowledge creation starts with socialization; tacit knowledge is created and shared through direct experiences with others or with the environment. When tacit knowledge is articulated through dialogue and reflection, it is externalized and converted into explicit knowledge. Explicit knowledge is collected, edited, and processed to form more complex explicit knowledge through combination. Explicit knowledge is shared and converted into tacit knowledge that is applied and used in practical situations through internalization (Nonaka & Toyama, 2003). Incorporating tacit knowledge into the knowledge creation theory created the delineation between knowledge and information (Nonaka & Von Krogh, 2009). Although, definitions in the literature of both knowledge and information are vague and ambiguous (Gourlay, 2006). The spiral process of the SECI model moves at the ontological level (individual and organization) and epistemological level (tacit and explicit) to amplify the creation of organizational knowledge by reusing the existing knowledge (Scurtu & Neamtu, 2015).

Ba refers to the shared context, physical or virtual, in which knowledge is created, shared, and put into practice through personal interactions and dialogue (Luhman & Cunlifffe, 2013). Knowledge is context-specific; it depends on the time, space, and relationship with others. In other words, it is situated action and needs a context to be interpreted (Nonaka & Toyama, 2003). Ba is a shared dynamic context that explains the potentialities that stimulate or hinder knowledge creation activities (Nonaka & Toyama, 2003). Ba can be thought of as the human operating environment in which project teams operate (Morris, 2013).

The SECI-Ba model was popular with practitioners but was criticized by scholars, who contended that not all tacit knowledge can be made explicit (Nonaka & Von Krogh, 2009), and the company loyalty and collaborative culture of Japanese firms do not translate well to the United States (Luhman & Cunlifffe, 2013). Nations such as the United States, with more individualist cultures tend to rely more heavily on documentation for knowledge sharing, whereas Asian collectivist cultures rely more on personal interaction and relationships, and thus have a greater propensity to share tacit knowledge (Wiewiora, Murphy, Trigunarsyah, & Brown, 2014).

Gourlay (2006) claimed the SECI model was flawed because the examples offered as evidence did not support proof, and could more simply be explained by learning by doing and designing new tasks. While Gourlay (2006) emphasized that Nonaka did not give scientific knowledge enough attention, he admitted that the concept of internalization, converting tacit knowledge into explicit knowledge, is difficult to illustrate.

Nonaka (1991) argued that Japanese firms are more successful at creativity and innovation in dynamic environments because they understand the value of tacit knowledge. Western companies think of knowledge as explicit, comprising of facts and reports that are quantifiable, and apply that knowledge to other quantifiable areas, such as return on investments, cost reductions, and increased efficiency. Japanese firms, Nonaka (1991) suggested, are more focused on insights, hunches, and intuitions because they typically have more commitment at the individual level.

Nonaka's (1994) knowledge creation theory was criticized by scholars who claimed it departed from Polanyi's (1966) concept of tacit and explicit knowledge. Polanyi (1966) argued that all explicit knowledge has a tacit component; thus, they are inherently inseparable. Overall, there is no such thing as purely tacit or purely explicit knowledge. Furthermore, Polanyi (1966) regarded knowledge as the process of knowing, rather than an object or asset (Gourlay, 2006). Moreover, Nonaka and Von Krogh (2009) disputed this criticism and explained that the theory agreed with Polanyi, stating explicit and tacit knowledge are not separate but rather exist along the same continuum. The continuum is useful for both scholars and practitioners.

History has shown that researchers need the choice of which end of the knowledge continuum to focus on to advance the knowledge-based view of the organization (Nonaka & Von Krogh, 2009). Not all knowledge can be made explicit (Nonaka & Von Krogh, 2009).

The SECI knowledge conversion process has been criticized for not adequately taking into account social practices (Nonaka & Von Krogh, 2009). Nonaka and Von Krogh (2009) acknowledged that more research is needed to understand the relationship between organization knowledge creation and organizational social practices.

Social and practice perspectives. One of the early developments of organizational learning came out of a debate of location and nature of learning; this resulted in a shift of an epistemology of possession to one of practice (Easterby-Smith et al., 2000). Organizational learning occurs mainly through the interactions between people. In fact, the learning organization, defined by Senge (1994), was criticized for not taking into account social aspects, such as the actions and practices of people that create learning processes (Caldwell, 2012). The social and practice perspectives of organizational learning include the situated learning theory and the structuration theory.

Situated learning. Situated learning theory was introduced by Lave and Wenger in 1991 (Jugdev & Mathur, 2013). Situated learning theory departs from the view of knowledge as an asset or resource; rather, situated learning considers knowledge as a process. During the engagement in changing processes of human activity, knowledge is shared by being incorporated into the practices and routines of the workplace (Luhman & Cunlifffe, 2013). Knowledge, whether tacit or explicit, is situated or context-dependent; it needs context to appear meaningful (Gourlay, 2006). Individuals share views from their own context and collaborate to create knowledge in dynamically linked processes so the organization can evolve (Nonaka & Toyama, 2003). Situated learning occurs during the interactions between people, practice, knowledge, and environment; it is focused more on practice rather than instruction (Lave & Wenger, 1991). Therefore, situated learning applies to project management.

Structuration theory. The many theories of knowledge management may have a basis in Gidden's 1984 structuration theory, which focuses on the structural properties of social systems (Timbrell, Delaney, Chan, Yue, & Gable, 2005). Timbrell, Delaney, Chan, Yue, and Gable (2005) analyzed 18 articles with knowledge management theories and compared them to themes of structuration theory. Their results show that knowledge processes are social processes and that key themes of structuration theory are useful for understanding these processes and therefore knowledge management theories.

Structuration theory addresses human agency and social structure as interactive and equally important. The relationship between structure and agency are mediated by social practices (Caldwell, 2012). Structuration theory claims that agents within a social system are influenced by the structures of that system and continuously recreate those structures through interaction (Timbrell et al., 2005). Behavior is not the consequence of structure or agency, but occurs from the simultaneous and continuous interaction of both (Bresnen, 2016). It is the continual interactions of individual agents with the social structure that defines and reproduces both agents and structure (Nonaka & Toyama, 2003). It is through these social practices and interactions that knowledge is created. Everyday social interactions involve situated activities of human agents and the structures around them.

Dynamic capabilities. The concept of dynamic capabilities was developed because of the belief that the resource-based view did not adequately account for how and why firms develop competitive advantages in dynamic environments where the manipulation of knowledge resources is critical (Eisenhardt & Martin, 2000). The resource-based theory combines knowledge creation with strategy; however, this view does not address the continuous building of resources through the interactions and contradictions with the environment (Nonaka & Toyama, 2003). The dimension of the resource-based perspective that includes skill acquisition, management of knowledge, and learning, has the greatest potential to contribute to strategy (Teece et al., 1997). Dynamic capability is an extension of the resource-based theory that states that the organization's resources, tangible and intangible, combined with its capabilities, determine competitive advantage (Breznik & Lahovnik, 2014; Luhman & Cunlifffe, 2013). The continual renewal of the resource base of a firm as the environment changes is the key to maintaining a competitive advantage (Breznik & Lahovnik, 2014). Dynamic capabilities expand this view of resources to the development and manipulation of capabilities created by those resources (Tzortzaki & Mihiotis, 2014).

Dynamic capabilities fall under two streams of literature. The first view, based on Teece, Pisano, and Shuen (1997) focused on organizational performance and strategy in a dynamically changing environment; this view concerns manipulating internal and external resources to deal with rapidly evolving technology, uncertain markets, and regulatory environments (Davies & Brady, 2016; Davies, Dodgson, & Gann, 2016). The second stream is based on Eisenhardt and Martin (2000), who identified two kinds of dynamic capabilities, simple routines and specific routines, based on the uncertainty of the environment (Davies & Brady, 2016; Davies et al., 2016).

Teece et al. (1997) introduced dynamic capabilities to determine what concepts allow firms to achieve and sustain a competitive advantage in environments of rapid technology change. Dynamic capabilities was defined "as the firm's ability to integrate, build, and reconfigure internal and external competences to address rapidly changing environments" (Teece et al., 1997, p. 516). The dynamic capabilities framework suggests that competitive advantage depends more on enhancing internal processes than defensive business strategies (Teece et al., 1997).

Dynamic capabilities combined organizational learning and knowledge management with effective business strategies (Krzakiewicz, 2013). The seminal paper by Teece et al. (1997) is one of the most influential articles in management studies in the 1990s, specifically in strategic management (Leybourne & Kennedy, 2015). Leybourne and Kennedy (2015) further elaborated on the 1997 definition of dynamic capabilities. Dynamic capabilities are high order capabilities in the form of intangible assets, that are valuable and difficult to duplicate by competitors, designed to change the resource base (renew/redeploy/reduce/destroy) to achieve and sustain competitive advantage, specifically in environments of rapid change (Leybourne & Kennedy, 2015).

Teece's (1997) dynamic capabilities theory differed from the resource-based theory since it added dynamic processes and coordination and combining of assets to create a competitive advantage, rather than relying on asset acquisition and development alone (Nonaka & Von Krogh, 2009). The high-technology industries in the global marketplace in the 1990s included semiconductors, information services, and software development. Firms, such as IBM, Texas Instruments, Phillips, and others, followed the resource-based strategy of accumulating valuable technology assets, but this approach is often not sufficient to maintain a significant competitive advantage (Teece et al., 1997).

To enhance and support internal and external competencies effectively, firms must have timely responsiveness, and rapid and flexible innovation capabilities to address rapidly changing environments (Teece et al., 1997). Dynamic capabilities are the firm's ability to develop and renew competences and strategically manage skills and resources in changing environments (Teece et al., 1997). Dynamic capabilities represent the organization's ability to rebuild its resource skills, and become innovative (Zollo & Winter, 2002). Organizational and managerial processes, such as assets and resources, shape competitive advantage, which combined, represent competences and capabilities, as well as the paths available to the firm (Teece et al., 1997).

The theory of dynamic capabilities is tautological when high performance is observed and attributed to resources; when dynamic capabilities are evaluated in regard to resource manipulation, independent of performance, this issue is avoided (Eisenhardt & Martin, 2000). However, firms can develop dynamic capabilities separately by incorporating best practices, the concept of inimitable, and non-substitutable does not apply, and therefore cannot result in a competitive advantage that is sustainable (Eisenhardt & Martin, 2000). This disagrees with the viewpoint of Teece et al. (1997) who stated that dynamic capabilities are unique to a firm. The definition of dynamic capabilities offered by Zollo and Winter (2002) concerned modifying processes and routines in pursuit of improved effectiveness, which is not tautological because in this definition, dynamic capabilities do not necessarily improve performance (Helfat et al., 2007).

Both dynamic capabilities and the resource-based view have been criticized for being vague and tautological, as reported in the literature review by Breznik and Lahovnik (2014). Eisenhardt and Martin (2000) reported that dynamic capabilities are not vague or tautological, but rather specific identifiable processes or routines (Eisenhardt & Martin, 2000). Dynamic capabilities have commonalities across firms, known as best practices, hence, they are more homogeneous than commonly assumed (Eisenhardt & Martin, 2000). In highly dynamic environments, dynamic capabilities are simple routines with unpredictable outcomes, that rely on newly created, situation-specific knowledge, and have an interactive execution (Eisenhardt & Martin, 2000). In moderately dynamic markets, dynamic capabilities resemble specific routines that rely on existing knowledge and linear execution with predictable results (Eisenhardt & Martin, 2000).

Dynamic capabilities is a term consistent with most concepts of knowledge management; the notion that the firm's ability to sustain competitive advantage is based on its capabilities to manage its knowledge assets (Krzakiewicz, 2013). The goal of dynamic capabilities is to gain competitive advantage (Teece et al., 1997; Zollo & Winter, 2002). Dynamic capabilities have been considered abilities or processes; the focus should be on the roles individuals play in creating, implementing, and renewing dynamic capabilities (Di Stefano et al., 2014). Table 1 provides an illustration of the number of citations of the sources this research is based on.

Table 1

Citations and Sources

Source	Number of Citations (From Google Scholar)
(Cangelosi & Dill, 1965)	651
(Polanyi, 1966)	24,367
(Senge, 1994)	46,289
(Teece et al., 1997)	25,730
(Eisenhardt & Martin, 2000)	11,379
(Zollo & Winter, 2002)	5,372
(Teece, 2007)	4,859
(Newell & Edelman, 2008)	62

Review of the Literature

Introduction

The topic of this dissertation is the relationship between knowledge transfer, team learning, and project success in the IT field. The theories presented previously include organizational learning and the resource-based view. Knowledge management and the development of dynamic capabilities extended organizational learning and the resource-based view. There is little research on dynamic capabilities as it relates to project management. Information technology includes complex changes and uncertainty and thus is an appropriate environment for dynamic capabilities. The literature review that follows will relate the previous theoretical concepts to the specific field of project management and IT. An analysis of the literature follows, organized by the variables used in this study. The research design used in this study is also addressed with the literature reviewed.

Project Management

Projects have been around since the beginning of time, but the concepts and characteristics of modern project management did not emerge until the 1950s (Morris, 2013). Professional project management societies were developed in the late 1960s and early 1970s (Morris, 2013). Project Management Institute published the first *Project Management Book of Knowledge* (PMBOK) in 1987. The *Project Management Book of Knowledge* was meant to cover knowledge unique to project management (Morris, 2013). The *Project Management Book of Knowledge* is so widely disseminated that it dominates the discipline, even though it rests on an implicit and narrow theory (Koskela & Howell, 2008; Morris, 2013). Koskela and Howell (2008) claimed that the lack of an explicit theory of project management leads directly to the problems in practical project management. "The improvisation and innovation critical to effective project management in complex environments is antithetic to the tenets of the PMBOK—which is widely accepted as documenting the tenets of good project management" (Leybourne & Kennedy, 2015, p. 22).

Abyad (2012) stated that a limitation of the *Project Management Book of Knowledge* is it does not address technology, environment, and business issues. Information technology project-based knowledge includes technology knowledge, organizational knowledge, and business value knowledge (Reich, Gemino, & Sauer, 2012). The *Project Management Book of Knowledge* has also been criticized because it does not address the effect of human behavior (Morris, 2013), or the idea that knowledge sharing through communication is a social phenomenon (Jackson & Klobas, 2008). Project management as a practice covers widely distinct industries and goals which provide a challenge to institutionalize the discipline; the possibility of creating a complete knowledge base regarding project management is questionable (Bresnen, 2016). Project management faces challenges from an individual, team, and environment perspective.

Project management presents challenges at the individual level because of the diversity of team members; it is at this level where organizational learning begins. Projects are uniquely challenged to share knowledge because of their temporary nature, diverse team, and innovative aspects (Holzmann, 2013). Project management is a social construct that is dependent on situational context, and values and beliefs (Jackson & Klobas, 2008; Morris, 2013). Process can only go so far, and then it comes down to people (Morris, 2013, p. 12). People are animate, they have passions, egos, ideas, values, emotions, and they make mistakes (Morris, 2013). Organizational learning begins with stresses at the individual level (Cangelosi & Dill, 1965). Project team members often experience conflicting pressures, especially if they are in a matrix organizational structure where they are members of various organization and identity groups (Chronéer & Backlund, 2015; Slepian, 2013).

Project management also has challenges at the team level and effective knowledge management is needed. Getting people to learn and to work effectively in complex, one-off situations, is challenging because it requires a careful balance of risk and reward (Morris, 2013). This situated discourse allows team members to compare conflicting perceptions and collaborate to solve problems (Easterby-Smith et al., 2000).

> Examples of activities used to generate practice-oriented learning include problem solving, asking for information, seeking experience or advice, reusing assets, coordinating, encouraging synergy, discussing developments, documenting experiences, visiting each other to learn by example, and mapping both knowledge and related gaps (Jugdev & Mathur, 2013, p. 640).

Projects are often self-organized; the project team is composed of cross-functional team members rather than as members of a traditional team based on a division of labor that follow traditional managerial practices (Cacciatori et al., 2011). Project teams are typically cross-functional and temporary, charged with delivering outcomes on tight schedules with limited budgets, and scope changes (Jugdev & Mathur, 2013). Conflicts are often escalated vertically to management rather than discussed in group meetings (Slepian, 2013). Learning in this sense may emphasize negotiating current relationships and changing relationships (Easterby-Smith et al., 2000). Projected environmental challenges include culture, experience, and size of the project team. Additional pressures result from project complexity and technology challenges (Jugdev & Mathur, 2013).

Knowledge Management in Project Management

Knowledge management is based on the belief that knowledge is crucial to an organization's competitive advantage (Krzakiewicz, 2013; Luhman & Cunlifffe, 2013; Nonaka, 1991). The purpose of knowledge management is to identify and transfer knowledge within the organization to improve business performance (Leybourne & Kennedy, 2015). This concept can be applied to IT projects that rely on knowledge sharing processes to resolve problems and collaboration to meet project goals. In extension, increased project performance should reduce the time, cost, and missed opportunities of working on failing projects, which may contribute to improved business performance.

Projects are dynamic; they create and define problems, develop and apply knowledge, and create new knowledge through action (Nonaka & Toyama, 2003). Situational learning theory is appropriate for project management because the practices of the discipline are experiential (Jugdev & Mathur, 2013). Jugdev and Mathur, (2013) believed situated learning theory could be leveraged in project management practice to improve both project learning and cross-project learning, the mediating variables used in this study. Through practice and routine, situated learning applies to the independent variables that constitute the construct of knowledge transfer. "All projects exist within a situated context and frequently exhibit what are often regarded as unique properties and contingencies" (Hall, Kutsch, & Partington, 2012, p. 682).

Research by Timbrell et al. (2005) illustrated at least 18 distinct theories of knowledge management. Gourlay (2006) stated that knowledge is consequences or components of behavior and suggested it can be managed indirectly through managing behavior. Knowledge management activities should add value to the organization; however, there is no standard for measuring that value (Scarso & Bolisani, 2011). This creates ambiguity in effectively managing the activities and behavior of individuals from a knowledge management perspective (Scarso & Bolisani, 2011). The competencies needed for knowledge management are unclear (Scarso & Bolisani, 2011). It is difficult for the organization to encourage knowledge sharing behaviors if their value cannot be adequately assessed and rewarded (Singh & Soltani, 2010).

Leybourne and Kennedy (2015) agreed with the behavior approach because of the importance of improvisation. Leybourne and Kennedy (2015) stated that improvisation happens when timetables are temporarily unattainable, or when complexities introduce ambiguity, which is common in project management. Innovation is a human behavior that allows a more flexible approach to project management than the plan then execute concept addressed in the *Project Management Book of Knowledge* (Leybourne & Kennedy, 2015). Flexibility in human behavior generates more innovative practices rather than just reproducing accepted practices (Leybourne & Kennedy, 2015). This concept is similar to the idea presented by Eisenhardt and Martin (2000) regarding dynamic capabilities in highly dynamic markets requiring simple routines that can be broadly applied and have an interactive execution.

Improvised behavior is not planned, but enacted by individuals that have gained a level of expertise and have tacit knowledge or intuition about the actions needed to find a solution (Leybourne & Kennedy, 2015). When individuals have tacit knowledge, they know the rules without having to look at the rulebook; they have developed the ability to act spontaneously and intuitively (Leybourne & Kennedy, 2015). These improvised activities may lead to developing best practices and innovation (Leybourne & Kennedy, 2015). The behaviors that support both project team learning and cross-project team learning will facilitate the creation and development of dynamic capabilities (Newell & Edelman, 2008).

Knowledge Sharing/Knowledge Transfer

The field of knowledge transfer in the project management field is rapidly developing, especially in the IT sector (Holzmann, 2013). Holzmann (2013) conducted a meta-analysis on 72 articles in the previous 10 years from 53 different journals. The articles were regarding knowledge brokering, and knowledge transfer. The author reported that the most studied industry sector was engineering. The second highest sector was IT because of the overall growth of the industry, and the relationship between information, knowledge, and IT projects (Holzmann, 2013).

Knowledge is the most important resource needed for project management (Gasik, 2011); and knowledge sharing is critical to project success (Holzmann, 2013). Knowledge transfer occurs when one individual or group seeks knowledge from another more experienced individual or group that results in changes of behavior of the first individual or group (Hall et al., 2012).

The transferability of knowledge depends on its tacitness; tacit knowledge is more difficult to transfer than explicit knowledge (Dinur, 2011). Both tacit and explicit knowledge are needed in IT projects. A similar analogy to the riding a bike example is reading the manual on how to build a server and actually building one. Likewise, there are many aspects of IT technical support gained through experience and skill and not from industry certifications. Knowledge can be managed by identifying what makes it tacit and then determining the most effective transfer method (Dinur, 2011). The most effective transfer channel depends on the type of knowledge being transferred; it also requires an active recipient who desires to learn (Dinur, 2011). Transfer channels can include codification, dialogue, practice, long term visits, apprenticeships, and social interactions (Dinur, 2011). Knowledge sharing is externalizing knowledge through communication by speech, artifacts, or gestures (Jackson & Klobas, 2008). Knowledge is transferred through individual experience, knowledge articulation, and knowledge codification.

Dynamic Capabilities in the Information Technology Field

IT project management is an appropriate environment in which to apply dynamic capabilities, because the framework partly addresses environments where there is rapid technology change (Teece et al., 1997). It is particularly difficult to deliver successful projects in IT, especially if they involve new, unproven technology (Morris, 2013). The dynamic environment in IT project management is influenced by the rapid changes in technology, and business goals, coupled with the often multiple IT solutions for the same business need (Paramkusham & Gordon, 2013). Project team members must use relationship building, decision making, and information sharing skills to achieve an optimal solution (Slepian, 2013). These uncertainties create a dynamic environment that requires continuous learning.

Breznik and Lahovnik (2014) recognized dynamic capabilities as a source of competitive advantage in the IT industry, where acquiring new knowledge and developing the resource base to apply that knowledge is a major factor for success. Dynamic capabilities can be applied to project management because they involve cooperation of cross-functional teams with different areas of knowledge (Eisenhardt & Martin, 2000). Dynamic capabilities applied to IT are defined as "a firm's ability to integrate, build, and reconfigure IT-enabled resources concurrently with organizational and managerial processes in order to align with a rapidly changing competitive environment" (Lim, Stratopoulos, & Wirjanto, 2012, p. 50). Capabilities have not been widely researched from a project management view (Rungi, 2014). Dynamic capabilities from a knowledge view were defined as "the capacity of an organization to purposefully create, extend or modify its knowledge-related resources, capabilities or routines to pursue improved effectiveness" (Salunke, Weerawardena, & McColl-Kennedy, 2011, p. 1252).

Davies and Brady (2016) presented the concept of project capabilities as dynamic capabilities that are operational rather than strategic, but have a reciprocal relationship that is mutually reinforcing since projects are a method of implementing strategy. Project capabilities refer to the knowledge, activities, and structures required to manage a project throughout its lifecycle, from conception through implementation to closure (Davies & Brady, 2016). Ahern, Byrne, and Leavy (2015) relate organizational learning, knowledge management, and dynamic capabilities to project capability development. Projects often have incomplete knowledge at the initiation of the project, therefore continuous learning and knowledge sharing are needed throughout the project lifecycle (Ahern, Byrne, & Leavy, 2015).

In 2007, Teece created a new classification of dynamic capabilities that include the capacity to sense new opportunities, the capacity to seize those opportunities, and the capacity to preserve competitiveness by maintaining, and reconfiguring tangible and intangible assets (Teece, 2007). This framework of sensing, seizing, and maintaining form "the beginnings of a general theory of strategic management in an open economy with innovation, outsourcing, and off-shoring" (Teece, 2007, p. 1347). The dynamic capabilities concept of sensing and seizing can be compared to those of Zollo and Winter (2002). Goal differentiation occurs through sensing and knowledge articulation; goal integration occurs through seizing and knowledge codification (Ahern et al., 2015).

While dynamic capabilities have proved useful for strategic management research and practice, it has little empirical support (Breznik & Lahovnik, 2014). The IT industry is currently the most dynamic environment, and IT firms influence other firms because of IT integration with other industries. Therefore, Breznik and Lahovnik (2014) choose the IT industry for their research. They interpreted the capacity of sensing as knowledge creation, the capacity of seizing as understanding knowledge value and acting on it, and reconfiguring as implementing change for knowledge management and integration. These were analyzed in several areas of the firms: managerial, marketing, technological, research and development, innovation, and human resource capability. In-depth interviews of six IT companies were conducted, which showed that developed and deployed dynamic capabilities are directly related to the firm performance; however, the resources must be continuously developed and adapted to fit the changing environment. Breznik and Lahovnik (2014) asserted that the three categories, sensing, seizing, and reconfiguring, are interdependent and all three must be developed and deployed; ignoring one can negatively impact the other two. Krzakiewicz (2013) agreed that dynamic capabilities place more prominence on change management over operational routines; changes in routines because of experience and knowledge sharing allow continual adaptation.

Naldi, Wikström, and Von Rimscha (2014) chose the European audiovisual production industry for their dynamic capabilities research because the industry is dynamic and it has a high need for innovation. The results of the survey research indicated that sensing and seizing capabilities enhance a company's innovative performance. This agrees with prior research that supports the observation that creative companies do not need additional creative skills, but rather need to develop and deploy sensing and seizing capabilities and the routines and processes that support them (Naldi, Wikström, & Von Rimscha, 2014). Implementing dynamic capabilities is costly (Breznik & Lahovnik, 2014), and somewhat difficult, otherwise the sustainable advantage would erode with the effective application of concepts of dynamic capabilities (Teece, 2007). This was empirically shown in the survey results by Naldi et al. (2014). They found that in the audiovisual companies, very low levels of sensing and seizing capabilities are not sufficient to influence innovation. An adequate amount of resources must be engaged before benefits can be realized (Naldi et al., 2014).

An alternative to Teece's definition of dynamic capabilities is "a learned and stable pattern of collective activity through which the organization systematically generates and modifies its operating routines in pursuit of improved effectiveness" (Zollo & Winter, 2002, p. 340). Building on the link between organizational learning and dynamic capabilities, Zollo and Winter (2002) defined three learning mechanisms that should contribute to the creation and development of dynamic capabilities, which in turn will enhance the evolution of organizational routines. The important question was to what degree and combination are these learning mechanisms the most effective to create and develop dynamic capabilities (Zollo & Winter, 2002). Following the example of Newell and Edelman (2008), the research method used in this study will focus on the three learning behaviors identified by Zollo and Winter (2002). The three learning mechanisms, individual knowledge (experience accumulation), knowledge articulation, and knowledge codification, make up the independent variables in this research and will be presented next.

Individual knowledge. The subconscious learning behavior (tacit knowledge) of the individual, based on experiences, was referred to by Zollo and Winter (2002) as experience accumulation. Experience accumulation will be referred to in this research as individual knowledge. This knowledge asset is the least costly for an organization, as opposed to knowledge articulation, and knowledge codification (Zollo & Winter, 2002). Individual knowledge is tacit knowledge an individual possesses from trial and error learning that influences current skills and behaviors. Individual knowledge relies upon memory, which is more easily recalled with increased frequency and repetition of the task involved (Zollo & Winter, 2002). Forgetfulness is common in projects because of their temporary nature, and being organized separate from other projects (Cacciatori et al., 2011). Individual knowledge is effectively shared by moving people within the organization, or hiring people based on personal experience (Mueller, 2015). Selecting the best project team is crucial to exploit individual knowledge.

Paramkusham and Gordon, (2013) found that in environments where expertise is insufficient or distributed among the project team, rather than in individuals, knowledge transfer is challenged because of fluid and ambiguous deliverables resulting in lack of time and bandwidth for team members. Time is the most potent resource in project management (Morris, 2013); and was found to be a limitation of knowledge transfer in all five case studies of project learning by Hartmann and Dorée (2015). By transferring individuals to other projects, the individuals will bring with them their knowledge, including tacit knowledge, which is difficult to transfer. The transfer of knowledge from one project team to another project team should result in the modification of operating routines associated with project work, therefore, it is an example of a dynamic capability (Newell &

Edelman, 2008). Learning and skill development from experience is more effective when tasks are similar. If tasks are not similar, or are not done frequently, knowledge articulation and knowledge codification will be more effective learning mechanisms (Zollo & Winter, 2002).

Knowledge articulation. Knowledge articulation allows the conversion of tacit knowledge to explicit knowledge (Nonaka, 1991). Dialogue and discussion allow team members to integrate individual perspectives into a collective perspective (Nonaka, 1991). Knowledge articulation includes team meetings, brainstorming sessions, casual conversations, and discussions with project team members, stakeholders, and end users. However, only a small percentage of articulable knowledge is actually articulated (Zollo & Winter, 2002). The potential of ungathered knowledge articulation could create a challenge for firms wishing to capitalize on this learning mechanism. Zollo and Winter (2002) pointed out that the process of codification may enhance individual knowledge. Therefore, the process of knowledge articulation may also enhance individual knowledge.

Knowledge codification. Knowledge codification involves codifying one's individual knowledge into electronic repositories that are available to peers, thus converting tacit knowledge into explicit knowledge. Knowledge is transferred from the individual to the organization through codification (Liu, Ray, & Whinston, 2010). Often individuals are offered rewards as incentives to transfer their knowledge in this manner.

Knowledge codification is the third learning mechanism defined by Zollo and Winter (2002). Knowledge codification is the most costly of the three learning behaviors for the firm to implement (Zollo & Winter, 2002). However, Gharaibeh (2012) reported that documentation is important to control project costs. Firms spend a lot of resources on knowledge management systems to capture and disseminate knowledge (Hall et al., 2012). The challenge for the firm here is determining the proper implementation because it could produce bad results if done poorly (Zollo & Winter, 2002). Codification includes explicit conclusions about actions taken; the codification process has benefit over just the output produced, as there is learning involved in participating in the codification as individuals reflect on practices and procedures (Cacciatori et al., 2011, Zollo & Winter, 2002). Knowledge codification in projects includes status reports, financial reports, work breakdown structure, project schedule, issue and resolution reports, risk mitigation documents, as well as electronic repositories, such as databases. A common document for knowledge codification is lessons learned.

Lessons learned, the learning gained during a project, is fluid; knowledge changes as it is shared (Jugdev, 2012). The *Project Management Book of Knowledge* includes several terms that refer to lessons learned, such as audits, project reviews, and best practices (Jugdev, 2012). Because of this ambiguity, researchers interested in learning gained during a project have no theory to base it on, therefore knowledge management theory and situated learning theory help to anchor the topic (Jugdev, 2012). Process knowledge is more difficult to share as it includes why something was done in addition to what was done, making it challenging, and more time consuming to codify (Newell et al., 2006).

Often organizations do not place enough emphasis on lessons learned and therefore a lot of project knowledge remains in tacit form as experience; but experience only benefits the individual if it is not shared (Leybourne & Kennedy, 2015). Knowledge independent of an individual or team must have a well-defined form and structure; this represents knowledge codification (Gasik, 2011). Codification efforts draw out explicit conclusions about the implications of actions and behaviors, something that articulation alone or experience alone does not do (Zollo & Winter, 2002).

Codification Versus Articulation

There is an additional challenge on managing knowledge because of differences in tacit and explicit knowledge; individuals may choose one method of knowledge transfer over the other. Formal rules for knowledge codification can be too strict and hinder knowledge sharing based on voluntary participation (Scarso & Bolisani, 2011). Liu, Ray, and Whinston (2010) pointed out that knowledge articulation through interpersonal networks and knowledge codification are interconnected. Collaboration is equivalent to personal networking (Tzortzaki & Mihiotis, 2014). Although the view that codification is a substitute for knowledge transfer through network ties is common (Cacciatori et al., 2011). Codification offers team members an option for knowledge sharing if they lose their network ties. As the benefits of codification increase, team members may choose to codify their knowledge over network sharing, weakening their network ties. However if network ties are strong, team members may refrain from codification to protect these ties; this is referred to as knowledge hoarding (Liu et al., 2010).

Sharing potential is determined by the frequency of knowledge seeking, and the perceived value of future knowledge sharing (Liu et al., 2010). Sharing potential is positively related to the expected tenure and the number of opportunities for developing network ties. The analysis of the game-theoretic framework developed by Liu et al. (2010) predicted that when the sharing potential is low, the tension between codification and articulation is high and people will choose one over the other, determined by the value of the codification reward. Likewise, when the sharing potential is high, the tension between codification and articulation is low; people will continue to use both forms of knowledge sharing. It is important to understand these interactions to determine a knowledge management approach that is the most beneficial (Liu et al., 2010). Formal practices, such as training, and informal practices as realized in organizational structure and culture, can increase opportunities to develop network ties (Mueller, 2015). However, this game-theoretic framework did not account for second hand knowledge and did not account for the fact that knowledge articulation through network ties may create additional knowledge, which may increase knowledge codification (Liu et al., 2010). Social interactions for knowledge sharing involve trust, and the acceptance of risk, because of the belief that the outcome will be positive (Wiewiora et al., 2014). When projects are going well, team members are confident, this removes the risk and eliminates the perceived need for trust-building relationships (Wiewiora et al., 2014).

Knowledge sharing through interpersonal networks of people allows the individual to retain their value as a knowledge source and is rewarded by peers through reciprocity (Liu et al., 2010). Organizational incentives may also be put in place. Network ties are more preferred in eastern cultures such as China and Japan (Davison, Ou, & Martinsons, 2013; Luhman & Cunlifffe, 2013). Knowledge sharing operates through individuals and implies a mutual obligation to reciprocate (Davison et al., 2013). Individuals choose to share knowledge-based on the rewards they expect to receive (Tzortzaki & Mihiotis, 2014; Zheng et al., 2011).

The behavior approach to organizational learning acknowledges that individuals and groups often fail to understand how deliberate knowledge sharing can influence project and organizational outcomes (Cangelosi & Dill, 1965). Newell and Edelman (2008) discovered that individuals within project teams perceived network ties to be more beneficial to knowledge sharing; however, survey results showed that codification was positively related to project learning, cross-project learning, and project success. Project team learning and cross-project learning are the mediating variables in this study and will be presented next, followed by project success, the dependent variable.

Project team learning. Project team learning is defined as the creation of knowledge within the project team that results in changed behaviors (Newell & Edelman, 2008). There are very few empirical studies completed on project team learning (Gharaibeh, 2011). However, team learning is critical for organizational learning (Gharaibeh, 2012). Successful project teams require communication and collaboration (Slepian, 2013). Collaboration is key for knowledge creation and sharing (Tzortzaki & Mihiotis, 2014); it is the shared goals of the project work that increase knowledge sharing to solve problems and uncover opportunities and challenges (Hartmann & Dorée, 2015). The project team is a group of individuals that do not share social ties to keep them focused, and communication channels are often weak (Gharaibeh, 2012). In a qualitative study of two projects Gharaibeh (2011) offered barriers to project team learning. Many of these barriers are the same as the barriers of individual learning, such as lack of time because of heavy workloads, lack of training, mentorship, and communication, which would increase opportunities for network ties. In addition, project team learning is challenged because of team members not understanding the similarities and dependencies between projects, and thus not intentional (Gharaibeh, 2011). Furthermore, Gharaibeh (2012) recognized that without project team learning, the ideal of the learning organization was not achievable.

Team learning increases when individuals who share the same interests and goals interact socially and exchange ideas and opinions (Gharaibeh, 2012). Project teams have a shared context and interact to create effective reflection and develop a collective perspective (Nonaka, 1991). Dialogue and discussion should include conflict and disagreement (Nonaka, 1991). Challenges to team learning occur when individuals and project team members work in isolation from other individuals and other projects (Gharaibeh, 2011; Paramkusham & Gordon, 2013). In addition, when team members are not co-located, team learning is challenged because of the difficulty of having socialization, brainstorming, and team building sessions (Gharaibeh, 2011). These sessions would foster relationship building and open discussions to enhance team learning (Gharaibeh, 2011).

Learning within the project team is critical since not all projects have complete knowledge and can be executed in a linear approach as traditional project management suggests. A level of risk and uncertainty is common in complex projects, and common in technology projects. When not all information is available up front, the project team must work together to learn from each other and to develop new knowledge throughout the project lifecycle (Ahern et al., 2015)

Cross-Project Learning

Cross-project learning is essential in IT projects, which are based on integrating rapidly changing technologies (Zhao, Zuo, & Deng, 2011). Cross-project team learning is defined as the transfer of knowledge within a project team to other project teams in the organization (Newell & Edelman, 2008; Zhao et al., 2011). There is a gap in the literature regarding cross-project knowledge sharing; more focus has been placed on the individual level and inter-organizational level (Mueller, 2015). The understanding of the transfer of knowledge across projects is limited (Zhao et al., 2011). Cross-project learning is critical to organizational learning, development and maintenance of competitive advantages, and innovation (Cacciatori et al., 2011; Holzmann, 2013; Mueller, 2015). Cross-project learning is challenging because of the structure of projects as separate goals from other projects; project team members must focus on knowledge sharing with other projects while maintaining separation in completing their own scope of work (Mueller, 2015). Project team members typically do not seek knowledge from other project teams unless there is a problem or issue to be resolved (Newell et al., 2006, Singh & Soltani, 2010). When the project is on target and perceived to be going well, project team members focus on the work at hand and do not spend time seeking knowledge that could be beneficial.

The nature of projects makes project team learning and cross-project team learning difficult to study from an objective perspective. Other barriers to cross-project learning are reluctance to blame or report failures, and the belief that the project to too unique, implying past project experience is not applicable (Gharaibeh, 2012; Hall et al., 2012; Newell & Edelman, 2008). An efficient practice to achieve cross-project learning is by transferring individuals between projects (Goffin & Koners, 2011).

A common practice for cross-project learning is documenting lessons learned and storing the information electronically, an example of knowledge codification. However, a study of 13 projects in six organizations found that the lessons learned information is often not accessed by other project teams, because of assumptions the knowledge is not beneficial, or a lack of awareness that the knowledge exists (Newell et al., 2006). Newell, Bresnen, Edelman, Scarbrough, and Swan (2006) discovered that the timing of lessons learned documentation was important. If lessons learned are completed too late, the project team may be already disbanded and team members do not feel it is important to contribute. If lessons learned are done too early, the successful outcome of the project may still be undetermined.

Landaeta (2008) suggested that cross-project learning could positively influence project success, but only up to a point. If a high level of effort is devoted to cross-project learning, the project performance may suffer because of increased costs and delayed schedule (Landaeta, 2008). Landaeta (2008) surveyed several large companies to test if cross-project learning increased project team learning and influenced project success. The regression analysis showed the level of knowledge transfer across projects was not a significant indicator of project success, although it was related to project team learning and project success (Landaeta, 2008). It was suggested that increased efforts in cross-project learning take away time from actually working the project and Landaeta (2008) suggested that the knowledge transfer tasks between projects be limited to one person. However, limitations to the study included the different levels of importance and learning capacities of individuals that responded to the survey (Landaeta, 2008). Implementation of one individual as a knowledge broker as suggested by Landaeta (2008), did not address the fact that individuals each have varying levels of skills and experience, and that tacit knowledge is difficult to transfer. Adding an additional person to the transfer method would hinder the amount and type of knowledge that could be transferred.

Hall, Kutsch, and Partington (2012) presented findings on two successive major projects in the United Kingdom government. The first one resulted in a highly publicized failure. In other words, it was highly articulated and codified, which are both needed for cross-project learning to take place. In addition, the knowledge sharing barriers of reluctance to place blame and report failure were eliminated. The second project had the same goal as the first; hence, they were not unique in their entirety. Consequently, due to the well-known failure of the first project, the second project team was more motivated to recognize risk and to learn from mistakes. The second project team was the active knowledge recipient described by Dinur (2011). Analysis of these two projects provided empirical evidence of how social and cultural factors influence the effectiveness of cross-project learning (Hall et al., 2012).

A case study of three Swedish project-based organizations in the engineering and construction field found that project members have a desire to learn and share knowledge but find it difficult because of time, resources, and system restrictions (Chronéer & Backlund, 2015). Meetings and experience exchange and feedback were found to be helpful for project team learning, although time constraints did exist and project team members felt there was much improvement to be made (Chronéer & Backlund, 2015). Cross-project learning was difficult as problems and issues encountered in the project were reported directly to senior management, but there was not a method for that information to be shared to other projects (Chronéer & Backlund, 2015).

Project success. Project success is a frequent topic in the project management literature but with different ideas on how to measure it (Abyad, 2012; Hadaya, Cassivi, & Chalabi, 2012; Pinto & Slevin, 1988b). A common method of measuring project success is related to budget, schedule, and performance, partly because these are controlled by the project team and are easy to quantitatively measure (Pinto & Slevin, 1988b). According to Morris (2013), many studies have stressed that projects should also be measured on effectiveness (meeting business goals). The knowledge areas with the highest influence on project success are time, risk, scope, and human resources (Abyad, 2012). However, there have been examples in the literature of projects that have come in over budget and behind schedule, and were still considered successful. The Sydney Opera House is one example. The project completed 93 million dollars over budget, and 10 years late, but is still considered a success because it is income-generating and globally famous (McKay, 2012).

Some projects considered successful may finish on time and under cost when they are not well received by the intended clients or end users. Accurately measuring client satisfaction can be challenging. Pinto and Slevin (1988a) recommended measuring project success using the three project criteria of budget, schedule, and performance, and adding three client criteria. The three client criteria used are how well the project's service or product works as it was intended to; satisfaction, how well the project's service or product's compatibility is with the needs of the clients; and effectiveness, how well the project's service or product contributes to the organization's effectiveness. Clients relate tacit knowledge back to the project team by their actions of using the product or service that was created or by not using it (Nonaka & Toyama, 2003). The survey developed by Newell and Edelman (2008) includes questions relating to both project criteria and client criteria to measure project success. The common concern is that for projects to be successful, they must be managed both effectively and efficiently, and are aligned to the organization's strategic objectives (Morris, 2013). Furthermore, the success criteria of a project varies depending on which phase the project is in, initiation, planning, execution, monitoring and control, or closing (Pinto & Slevin, 1988a). This leads to the conclusion that to compare project success rates, one should consider projects in the same phase. This research will only consider projects that are completed.

Brady and Davies (2014) performed a qualitative case study of two construction mega-projects. A key finding of their interviews from the Heathrow Terminal 5, and the London 2012 Olympic Park projects indicate that collaborative behaviors along with a flexible situational approach lead to both projects coming in on time and under budget. Integrated teaming was implemented in the Terminal 5 project; teams were assigned for each subproject and composed of the most talented resources (Brady & Davies, 2014). This supports the importance of individual knowledge in a large complex project environment. The London 2012 Olympic Park project implemented a rigid system of monitoring (reports and briefings), consisting of upward reporting and downward assurance (Brady & Davies, 2014). This supports the importance of knowledge articulation and knowledge codification in a large complex project environment. This is noteworthy, even though the case study was limited to two projects, because previous research suggests project complexity may be a challenge to project success. Furthermore, in both mega-projects the specific dynamic capabilities were developed through three phases: a learning phase acknowledges individual and team learning and experience; a codifying phase captures what was learned through articulation and documentation; and a mobilizing phase expands on current knowledge and modifies it to adapt and create new knowledge (Davies et al., 2016).

Mahaney and Lederer (2011) suggested that monitoring of IT project team members will allow the project manager to identify potential issues and make adjustments to bring the project back on target for successful completion. The study found that more monitoring of the activities leads to less privately held information that led to more project success; moreover, more monitoring will lead directly to more project success (Mahaney & Lederer, 2011). Monitoring includes such things as project team meetings, project reviews, codification of project progress, and post completion audits. Privately held information includes project knowledge, and project problem and issue status (Mahaney & Lederer, 2011). These findings substantiate the research done by Ahern et al. (2015) on two complex projects in two Irish utility companies. *Railco* and *Energyco* developed complex project capabilities through continual learning throughout the project, by frequent formal and informal communication between team members (Ahern et al., 2015). Complexity is measured by the number of project team members, and determining how similar the project was to previous projects.

Research Design

Several authors in the field of project management have based their work on the resource-based view (Hadaya et al., 2012). A Delphi study asked both scholars and practitioners of IT project management to prioritize resources and capabilities, both panels placed capabilities in the top 10 list of importance (Hadaya et al., 2012). This agrees with previous research that concludes that how a firm manipulates its resources to create capabilities is more important than the resources themselves (Hadaya et al., 2012). Therefore, dynamic capabilities are a suitable theory for project management research in the IT field. Past research of dynamic capabilities was both quantitative (surveys) and qualitative (interviews) (Eriksson, 2013). In the literature review on dynamic capabilities research, Eriksson (2013) mentioned five studies with "advanced survey instruments" (p. 314), one of which was the survey developed by Newell and Edelman (2008), which was chosen for this study because it is focused on project management.

This research used structural equation modeling (SEM) analysis. Structural equation modeling is a common method of quantitative research when testing a model. Structural equation modeling and confirmatory factor analysis was used to test the multidimensional model used to develop an instrument to operationalize specific resource-based view concepts (Karimi & Walter, 2015). Hermano and Martin-Cruz (2014) used structural equation modeling to test a model of top management involvement in dynamic capabilities leading to project and portfolio performance. Zheng, Zhang, and Du (2011) followed the learning mechanisms of Zollo and Winter (2002) and considered dynamic capabilities as being knowledge-based and consisting of creating, acquiring, and integrating knowledge. Zheng et al. (2011) used structural equation modeling to test their model of network embeddedness, knowledge-based dynamic capabilities (acquisition, generation, and combination) and innovation performance. They removed the less significant results to adjust their model to one that better fit the data results. Mahaney and Lederer (2011) used structural equation modeling to test the model of contract type, goal conflict, monitoring, and privately held information on project success in information systems development projects.

The research conducted by Lichtenthaler (2009) used structural equation modeling analysis to understand different types of learning (exploratory, transformative, exploitative), to absorptive capacity and the influence on innovation and performance of the firm. Although the instrument used by Lichtenthaler (2009) was also considered an "advanced survey instrument" (Eriksson, 2013, p. 314), it was not chosen for this research because of the focus on absorptive capacity that considers knowledge gained from external sources. Many IT projects will have external sources, such as vendors or contractors, but the focus of this research was to treat the project members as a collective unit and not distinguish between internal and external resources.

Newell and Edelman (2008) based their research on a large utility company in the United Kingdom, because of the dynamic environment in which it was operating. Both interview and survey data were obtained. Interview data was based on two projects nearing completion, survey data was based on 400 ongoing projects. Structural equation modeling was used to test the quantitative data from the survey responses against the model. Survey results show that codification facilitates learning across project teams, which supports the development of an organization's dynamic capabilities.

Although there was theoretical development of dynamic capabilities and organizational learning, there have been few empirical studies focused on learning and project success (Newell & Edelman, 2008). This finding reported by Newell and Edelman (2008) supported the work by Zollo and Winter (2002) that knowledge articulation and knowledge documentation can stimulate dynamic capabilitiy development (Newell & Edelman, 2008). The intent of this study was to be able to compare results gained from IT project management in the United States to the results reported by Newell and Edelman (2008).

Findings

Project management theory is underdeveloped; project management research has borrowed the theories of other disciplines. Organizational learning is a common base theory in project management research. However, there is a need to develop a theory of project management that combines the static knowledge and the dynamic knowledge of projects (Ahern Leavy, & Byrne, 2014).

Very few studies existed that related dynamic capabilities to project management. Because of an interest in learning and the experience of the shortfalls in the project management profession regarding lessons learned, the goal was to research learning in IT projects. The scarcity of articles that apply dynamic capabilities to project management may be because the initial focus of dynamic capabilities, and the resource-based view on which they were based, was on strategic management. Organizations implement changes based on overall strategic decisions, and those changes are implemented with projects. It therefore seems to reason that dynamic capabilities at the organizational level can be compared to dynamic capabilities at the project level.

Other studies in dynamic capabilities have referenced the behavior theory introduced by Cyert and March in 1963 (Danneels, 2008). Ahern et al. (2015) stated that 90% of project failures could be attributed to people; however, no citation for this fact was reported. In the study by Davies and Brady, (2016), dynamic capabilities was referenced with the evolutionary theory. Indeed Zollo and Winter (2002) stated their three learning mechanisms in the context of evolving the dynamic capabilities of the firm. This research focuses on completed projects, which are a point in time. A limitation of this research is the current development of additional capabilities, in knowledge sharing and team learning, will not be addressed. The goal was to understand the influence on project performance to provide a guide for the management of future projects.

Ahern et al. (2015) referenced knowledge-based view and dynamic capabilities and implied the theme of organizational learning, that learning occurs at different levels of the organization, through problem solving. The research by Ahern et al. (2015) also linked project capability development with Teece (2007) and Zollo and Winter (2002).

Mahaney and Lederer (2011) used agency theory to explain project success. Agency theory focuses on the relationship between agents (project team members) and principals (organizations). Mahaney and Lederer (2011) indicated that agents might withhold knowledge and act in their own interest. The contract type between the agent and the principal may provide rewards that lessen the tendency to withhold information, and align goals. Contract types are not a standard project management activity and therefore agency theory was not addressed in this research. Agency theory considers organizational incentives to align ownership with control (Teece, 2007) which can occur between managers and the organization (Hermano & Martin-Cruz, 2014). The focus of this study is not on specific members of a project but rather the knowledge transfer activities occurring in and between projects. In their view of dynamic capabilities, Eisenhardt and Martin (2000) stated that in moderately dynamic environments, there is a strong reliance of existing knowledge and experience. In these environments tacit knowledge is codified (Eisenhardt & Martin, 2000), but the challenges other authors have put forward in codifying tacit knowledge was not addressed.

A clear need exists for more understanding of learning capabilities in project management. A strength of this study is that it used one of the few quality instruments that addresses the influence of knowledge transfer on project success. By focusing on knowledge transfer in the IT field, during the conduction of this study a comparison of results from the previous study by Newell and Edelman (2008) might also relate to projects in a different field. The comparison may uncover similarities that may be consistent in projects in general, and differences that may be due to the nature of the IT field. In either case, future research will be identified that may help to advance the practice of project management and contribute to theory development in the discipline.

Critique of Previous Research Methods

The research by Hall et al. (2012) found that highly publicizing the failures of one specific project allowed the knowledge gained during that project to be applied to a second similar project which was successful. However, the case study used by Hall et al. (2012) was limited to two projects and cannot be generalized. Survey results provided by Naldi et al. (2014) may indicate that complex projects which have a higher number of resources may be more successful at innovation at the organizational level. Limitations to their study in the audiovisual field included low response rate, and a single respondent per company (Naldi et al., 2014). In addition, the survey instrument could be developed further because of the difficulty of measuring innovation.

Rungi (2014) surveyed 189 companies across various industries in Estonia and found that project related capabilities are more significant to firm performance than business capabilities. A concern with the survey used by Rungi (2014), is that the measurement of project success was limited to a schedule, scope, and cost perspective, as is a common approach. The expectation is that the results from that study would identify additional quality criteria for project success. An interesting finding, however, is that project teamwork had a negative effect on all three measures for project success (Rungi, 2014). It takes time to build trust among team members and it can be speculated that perhaps shorter duration projects do not have the time to develop the social aspects of a well performing team. Rungi (2014) suggested that the negative impact of project teamwork might be because "engineers and scientists do not value teambuilding enough" (p. 252). This is a complicated issue and one would not expect teamwork in IT projects to vary across industries. It is also unclear if companies in Estonia can be compared to the United States because of cultural or managerial differences.

Research by Ahern et al. (2005) on complex project capability is limited based on the analysis of only two organizations, and cannot be generalized to a wider population without additional research. Fifty one interviews were conducted over 4 years with individuals at different levels of the project structure (Ahern et al., 2015). This was a concept paper and very little information was given regarding the research.

Research on knowledge-based dynamic capabilities was conducted on 218 Chinese manufacturing firms to understand the influence on innovation performance (Zheng et al., 2011). Although that study was not specific to project management, there should be some similarities as projects undertake goals that result in a product or service. Variables such as trust, joint problem solving, and commitment were included and were shown to be significant to knowledge acquisition. Trust, joint problem solving, and commitment are important to project teams as well. Zheng et al. (2011) also found that the diversity of knowledge assets was significant to knowledge combination. Which would indicate the benefit of cross-functional teams that specialize in different areas help to integrate and use the knowledge available in projects. Manufacturing firms was the population of the study but generalizing difficulties arise because of the diversity of manufacturing firms, such as chip manufacturers, and steel manufacturers (Zheng et al., 2011).

The mixed method approach used by Newell and Edelman (2008) was a strength of their study. The mixed method approach allowed for the comparison of survey responses with interview responses. Twenty one interviews were conducted on two complex projects that were close to completion. Interview questions were based on the specific projects and on projects in general at the organization. The survey data was from 400 ongoing projects. A limitation, not stated in the article, is that it may be difficult to determine project success if the project has not completed. Another strength is the instrument was one of the few advanced instruments in dynamic capabilities (Eriksson, 2013).

Hadaya, Cassivi, and Chalabi (2012) used the Delphi technique to gather information from scholars and practitioners. The Delphi method was appropriate for their research to obtain judgment from experts and come to a consensus on the level of priority of resources and capabilities. This was useful in their study because they asked to prioritize resources and capabilities that were most important in IT project management. This research is quantitative and aims to obtain perceptions of experiences in a completed project. Using the Delphi method for a similar study to this may be useful for a large project that would warrant further detailed investigation.

Landaeta (2008) conducted a mixed method research to gather data from 46 closed projects to understand the relationship between knowledge transfer across projects, the body of knowledge within a project, and project performance. The mixed method approach included surveys, a focus group, and interviews. The sample was large organizations in several countries. This would not account for some of the culture differences that influence knowledge transfer. Landaeta (2008) suggested a high knowledge transfer will take costs and resources away from the project but did not show this in the analysis.

Summary

The goal of this study was to gain an understanding of the influence of knowledge transfer in IT projects that contributes to project success. There is no overarching theory of project management or knowledge management because these fields cross several disciplines. Many research articles relating to project management have been based on organizational learning theory, which was initially established in the early 1960s. The evolution of organizational learning has created three streams of thought. The resource-based view later combined with knowledge management into the knowledge-based view is the first stream. The second stream focuses on the social and practice aspects that includes structuration theory and situated learning. The last stream is the development of dynamic capabilities.

Organizational learning, knowledge management, and dynamic capabilities can be applied to project management in the IT field because of the learning activities of projects and the dynamic nature of the IT environment. Independent variables used in this study include individual knowledge, knowledge articulation, and knowledge codification. Mediating variables in this study are project learning and cross-project learning, and the dependent variable is project success. These variables were presented in relation to the theories and research articles were compared and contrasted.

This research is based on the dynamic capabilities as outlined by Zollo and Winter (2002) and operationalized by Newell and Edelman (2008). The survey instrument is regarded as one of the advanced instruments in dynamic capabilities and was previously used on project management. This research will focus on project management in the IT field. The methodology used in this research will be presented in the next chapter.

CHAPTER 3. METHODOLOGY

Introduction

Chapter 3 will focus on the methods and procedures for this research study. This chapter begins by re-familiarizing the reader with the material that was briefly presented in Chapter 1. Sections of this chapter include the purpose of the study, the research questions and hypotheses, the research design, the target population, and the sampling procedures. The survey instrument used in this study will be presented along with the variables used. This chapter will close with the ethical considerations of this research.

Purpose of the Study

The purpose of this study was to gain more understanding about knowledge management as it relates to project management in the IT field. The field of project management is growing and IT projects are conducted in a diverse number of industries (PMI, 2013; The Standish Group, 2013). Industries that are beginning to expand their technology capabilities and using IT projects at an increasing frequency, may not have the extent of IT project management knowledge as more technology driven industries. Knowledge is the most important project management resource, and knowledge management is required for effectively managing projects (Gasik, 2011). It is important to understand how knowledge management in IT projects can be used to help increase project effectiveness across industries.

This study topic relates the knowledge management concept of knowledge transfer to dynamic capabilities as categorized by Zollo and Winter (2002). Both tacit and explicit knowledge are transferred through individual knowledge (experience accumulation), knowledge articulation, and knowledge codification (Zollo & Winter, 2002). This regression survey research will test the model of knowledge transfer, team learning, and project success controlling for projects in the IT field.

Conducting this study is important to project management practitioners in all industries because improving project success rates will help reduce wasted money and missed opportunities. Effective project learning can reduce repetitive mistakes and contribute to business success (Jugdev & Mathur, 2013). Project management has been around for a long time but there is no overarching theory of project management. The *Project Management Book of Knowledge* is the leading guide for project management practices in the United States. However, scholars have pointed out many limitations of the PMBOK (Abyad, 2012; Gasik, 2011; Koskela & Howell, 2008; Morris, 2013). The findings from this study may help encourage the suggestion that knowledge management receives more attention in the practice of project management. Thus, this study is relevant to both practitioners and scholars.

There is a strong interest in knowledge management from scholars. As stated previously (see p. 16), peer-reviewed articles in scholarly journals have steadily increased in the area of knowledge management, with an average of 261 articles per year between 2010 and 2015; 245 articles per year between 2006 and 2010; and 125 per year between 2001 and 2005. The concept of dynamic capabilities is also growing in popularity in research (Di Stefano et al., 2014), and being applied to project management. The increased trend and number of published articles shows there is a strong interest in this area. The findings from this study will build on previous research and apply it to the field of IT. The results should help show how knowledge transfer methods impact project success, and if there is any difference from previous research that was not specifically applied to IT projects.

Research Questions and Hypotheses

How does the model of knowledge transfer, team learning, and project success explain the relationship between project success and individual knowledge, knowledge articulation, and knowledge codification, controlling for the effects of IT projects? A regression analysis was used to test the overall hypothesis and determine the degree of the relationship between the variables.

RES Q1: What is the relationship, if any, between individual knowledge and project learning?

$H1_0$: There is no significant relationship between individual knowledge and project learning.

$H1_A$: There is a significant relationship between individual knowledge and project learning.

RES Q2: What is the relationship, if any, between individual knowledge and cross-project learning?

$H2_0$: There is no significant relationship between individual knowledge and cross-project learning.

$H2_A$: There is a significant relationship between individual knowledge and cross-project learning.

RES Q3: What is the relationship, if any, between knowledge articulation and project learning?

$H3_0$: There is no significant relationship between knowledge articulation and project learning.

$H3_A$: There is a significant relationship between knowledge articulation and project learning.

RES Q4: What is the relationship, if any, between knowledge articulation and cross-project learning?

$H4_0$: There is no significant relationship between knowledge articulation and cross-project learning.

$H4_A$: There is a significant relationship between knowledge articulation and cross-project learning.

RES Q5: What is the relationship, if any, between knowledge codification and project learning?

$H5_0$: There is no significant relationship between knowledge codification and project learning.

$H5_A$: There is a significant relationship between knowledge codification and project learning.

RES Q6: What is the relationship, if any, between knowledge codification and cross-project learning?

$H6_0$: There is no significant relationship between knowledge codification and cross-project learning.

$H6_A$: There is a significant relationship between knowledge codification and cross-project learning.

RES Q7: What is the relationship, if any, between project learning and project success?

$H7_0$: There is no significant relationship between project learning and project success.

$H7_A$: There is a significant relationship between project learning and project success.

RES Q8: What is the relationship, if any, between cross-project learning and project success?

$H8_0$: There is no significant relationship between cross-project learning and project success.

$H8_A$: There is a significant relationship between cross-project learning and project success.

Research Design

A survey design approach helped to obtain data that had not been altered by the presence of the researcher during the data collection for this study. An online-distributed survey is a common method of non-experimental research and can yield a large sample. A survey design was chosen because this study is, in part, replicating a previous study (Newell & Edelman, 2008) but targeting a different population. Newell and Edelman (2008) focused their study of dynamic capabilities in project management by surveying and interviewing individuals in a utility company in the United Kingdom. This study used an existing survey to collect data from IT project professionals in the United States across industries. Permission was obtained from both Dr. Newell and Dr. Edelman to use their survey in this research.

SurveyMonkey and Qualtrics are two popular companies that provide the capability to create surveys online and publish them, or have the companies send them to their audience members. Project Management Institute and the PMI LinkedIn group were initially contacted to distribute the survey created in SurveyMonkey. However, this approach did not yield many results, as there was not an active push to encourage people to participate. Qualtrics was then chosen to distribute the survey in this study because the combination of project management professionals, IT projects, and the United States was very small with SurveyMonkey. The strengths of this study include the random sampling method, the sample size, the survey instrument, and the non-interference by the researcher. In addition, comparing the results from this study to previous studies may help identify areas requiring further analysis. The survey instrument consisted of questions that were used to measure the variables in the research questions.

There are three main constructs in this research: knowledge transfer, team learning, and project success (see Figure 1). These constructs will be presented next, along with an explanation of the variables making up the constructs. The construct of knowledge transfer is the independent variable. Knowledge transfer relies on the closely related construct of knowledge sharing, which includes the willingness of the individual or team to share (Islam, Low, & Rahman, 2012). Knowledge transfer is the interaction between a sender and a recipient (Gasik, 2011). Knowledge transfer occurs when a recipient accesses knowledge that was relayed by personal experiences, discussed, or documented, and is expected for effective performance (Islam et al., 2012).

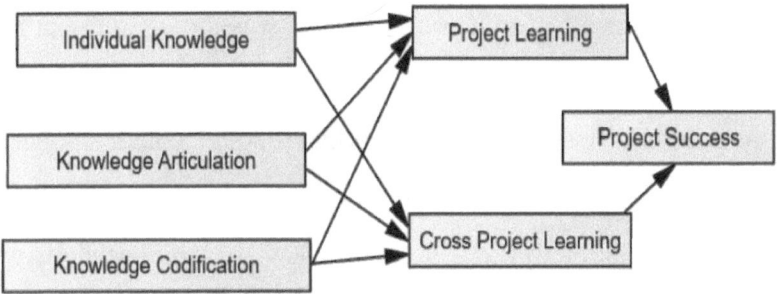

Figure 1. Structural model with arrows depicting the hypothesis relationships between the independent variables of individual knowledge, knowledge articulation, and knowledge codification, as they impact the mediating variables of project learning and cross-project learning, which should influence project success.

The following variable definitions were used for knowledge transfer. Three independent variables constitute this construct, individual knowledge, knowledge articulation, and knowledge codification. Individual knowledge is defined as the knowledge at the individual level resulting from values, experience, and capacity to learn (Goffin & Koners, 2011; Newell & Edelman, 2008). Knowledge articulation is knowledge that is shared through meetings and conversations. Knowledge articulation is particularly important for transferring tacit knowledge. Tacit knowledge is commonly shared by telling stories that individuals can learn from by relating them to their own experiences, but are difficult to document (Goffin & Koners, 2011).

Knowledge articulation often occurs during project phase gates and at project completion in meetings where the project team discusses lessons learned (Jugdev, 2012; Newell & Edelman, 2008). Knowledge codification is knowledge that is documented. Knowledge codification is specifically writing down what was learned (Lierni & Ribière, 2008; Newell & Edelman, 2008), or why decisions were made. Codified knowledge can be stored in a database for future project members to peruse. Lierni and Ribière (2008) stated that knowledge codification represents explicit knowledge. Project meetings, project reviews, and project close audits can be used to increase knowledge sharing and the potential for project success (Mahaney & Lederer, 2011).

The construct of team learning is the mediating variable. This is closely related to the theory of organizational learning. However, Caldwell (2012) stated that learning is done by individuals and that caution should be exercised when discussing organizational learning. In addition, Crossan, Maurer, and White (2011) explained that there is no overall agreement on what constitutes a theory, and an accepted theory of organizational learning has not yet been developed. Team learning has a basis in the theory of knowledge reuse, which considers the needs of the knowledge re-user. The needs of the knowledge re-user may be different from the anticipated needs of the individuals who created the knowledge archives, leading to rework (Markus, 2001). Team learning is the readiness of team members to effective use the collectively held knowledge (Jetu & Riedl, 2012).

The following variable definitions were used for team learning. Two mediating variables constitute this construct. Project learning is defined as learning within the project team. Cross-project learning is defined as learning between project teams. Situational learning theory is a view of workplace learning that focuses on learning between and within projects (Jugdev, 2012).

The construct of project success is the dependent variable. The following variable definitions were used for project success. One dependent variable constitutes this construct. Project success is defined as whether the project met the business needs and stayed within the business constraints. Landaeta (2008) measured success on cost, schedule, and quality. Lierni and Ribière (2008) measured project success on meeting schedule, budget, technical requirements, and fulfilling organizational criteria.

The model presented in Figure 1 represents a confirmatory factor model in which structural equation modeling analysis is appropriate (Blunch, 2013). Structural equation modeling can be used to measure the effects of latent variables on the dependent variables. In other words, structural equation modeling will test how well the data collected fit the model and whether the model should be further tested or altered. Structural equation modeling analysis does not prove causality. There is no expectation of proving that knowledge transfer impacts project performance in the conduction of this study.

Target Population and Sample

The sample is a subset of the target population that was studied. Samples are commonly used because obtaining data for the target population is usually not feasible. If done correctly, the sample of the target population should be random and representative of the target population. In this manner, results gathered from the sample can be generalized to the target population. This section will describe the target population, how the sample was chosen, and how the sample size was calculated so that the results could be generalized to the target population.

Population

Trochim and Donnelly (2006) differentiated between the theoretical population and the target population. Cooper and Schnider (2011) identified the target population as the people that hold the information and can answer the measurement questions. In this study, the theoretical population is IT projects in the United States; results of this study will be generalized to the theoretical population. Since is it not feasible to survey the theoretical population, a target population must be identified. The target population in this study is the project professionals in the United States who have worked on IT projects.

The Project Management Talent Gap Report stated that there are over 700,000 members of PMI (PMI, 2013). Moreover, the report claimed a growth of 15.7 million new project management roles would be created globally between 2010 and 2020. This was broken down to more than 2.3 million new project management jobs in the United States. The major project intensive industries include manufacturing, business services, finance and insurance, oil and gas, information services, construction, and utilities. Technology projects can span all these industries because IT products and services are encountered in a variety of situations, which is only expected to increase.

Project Management Institute is the "world's leading not-for-profit professional membership association for the project, program and portfolio management profession" (PMI, 2015, p. 6). The annual salary survey conducted by PMI reports demographics of the 9,677 survey respondents from the United States (see Table 2). These demographics can be generalized to the population of project professionals in the United States.

Table 2

Demographic Characteristics

Demographics	Population (from PMI)	Sample (used in this study)
Gender		
Male	62%	57%
Female	38%	43%
Experience		
< 5 years	12%	13%
5-10 years	29%	40%
10-15 years	24%	20%
> 15 years	35%	28%
Team Size		
< 10	51%	27%
10-20 years	30%	
11-25 years		25%
> 20	18%	
> 25		49%
Industries		
IT	21%	51%
Government	10%	4%
Healthcare	9%	3%
Consulting	8%	
Financial Services	7%	9%
Engineering	6%	
Manufacturing	6%	3%
Professional Services		9%
Energy & Utilities		7%

The population consists of 62% male, 38% female; 12% with less than 5 years' experience in project management, 29% with 5-10 years, 24% with 10-15 years, and 35% with over 15 years' experience. Education levels vary in the population; 11% have an educational degree in project management, 89% do not. Sixty percent of respondents reported they had worked on a project in IT. Project team size was fewer than 10 people for 51%, between 10 and 20 people for 30%, and more than 20 people for 18%. More than 20 industries were represented, including IT (21%), government (10%), healthcare (9%), consulting (8%), financial services (7%), and engineering and manufacturing (both at 6%).

The population characteristics reported by PMI (2015) are compared to the sample characteristics in this study in Table 2. It is important to note that the sample in this study was limited to those individuals that had worked on an IT project in the past year; the demographics provided by PMI do not have that limitation.

Sample

The sampling method used in this study was a random sample generated from Qualtrics, who distributed the survey through GlobalTestMarket. The target population is the subset of the GlobalTestMarket subscribers that have recently worked on an IT project in the United States. The target population is not limited by industry, only by project type being in IT. The exact number is unknown. A sample is a portion of the target population that is examined (Cooper & Schnider, 2011). The sampling frame is the method to access the target population (Trochim & Donnelly, 2006). In this study, the sample frame was the potential GlobalTestMarket subscribers. Participants who met the selection characteristics and completed the survey comprise the subsample (Trochim & Donnelly, 2006). Participant characteristics were individuals who had worked on an IT project in the United States. This yielded a large sample that, although limited to IT projects, spans across industries and includes projects of various size and complexity.

Coverage error can occur when the sample frame does not include all members of the population (Dillman, Smyth, & Christian, 2009). Since the potential GlobalTestMarket subscribers and the size of the total population are both large, coverage error in this study was minimal. Internet surveys have coverage error when not all members of the population have internet access. Since the population in this example includes individuals that work in IT, the coverage error for those without internet access is minimal to nonexistent.

Power Analysis

It is the sample size, and not the proportion of the population sampled, that affects precision (Dillman et al., 2009). Smaller sample sizes can be acceptable if the population is limited or the model is simple. In this research, the model is simple, but the population is large. Vogt (2007) stated that bigger samples are better because they increase statistical power and reduce sampling error.

Vogt (2007) explained that often sample size calculations are difficult because they are based on information that has not yet been obtained. The variables used to determine the sample size include the confidence level, the confidence interval (CI), and the split. The confidence level is 95%, this represents how likely the answer will lie between the confidence interval. The confidence interval is the range around a statistic, which is typically set as plus or minus five, this is often referred to as the margin of error. The split is the percentage of an answer that would vary. Fifty percent is the most conservative value of split. A 50/50 split means that 50% of respondents to a yes or no question would answer no and 50% would answer yes.

The structural relationships between the variables in a structural equation modeling model are regression equations, which represent the influence of one or more variables on another variable (Byrne, 2010). This influence, or relationship, is shown in structural equation modeling by arrows. Using the G*Power calculator (Faul, Erdfelder, Buchner, & Lang, 2009) to determine sample size for a multiple regression with margin of error of 0.05, confidence interval of 0.95, and an effect size of 0.2 returned a minimum sample size of 90. A guideline provided in the literature for sample size uses the $N:q$ rule, where N is the ratio of cases and q is the number of parameters in the model that require statistical estimates (Kline, 2011). This rule uses the ratio of 20:1 for an ideal minimum sample size (Jackson, 2003). In the model used in this study, six parameters, or variables require data analysis. Therefore, a sample of 120 for this model is considered an adequate sample.

Procedures

This section will provide details about the procedures followed to carry out this research. The information included in this section is the selection process, and assurance of protection of the survey participants. This will be followed by details on how the data was collected and analyzed.

Participant Selection

The qualifying criterion for this research includes individuals that have worked on an IT project in the United States. After the survey was created in Qualtrics, a link to the survey was e-mailed out to the subscribed audience of GlobalTestMarket. Individuals willing to take the survey clicked on a link in the e-mail. The first question of the survey asked, *Which of the following work related activities have you completed in the United States in the past year?* This was a multiple-choice question with four responses offered. If participants selected the answer *Worked on an IT project,* they were advanced to the next screen, which included the informed consent.

Protection of Participants

Three guidelines identified by Cooper and Schindler (2011) to protect the rights of the participants include explaining the benefits of the study, explaining their rights and protections, and obtaining informed consent (p. 32). Following recommendations by Creswell (2009), the initial communication to the qualified participants did the following: identify the researcher, state that this is for a dissertation at Capella University, describe the purpose and benefits of the research, explain the participant involvement, guarantee confidentiality and anonymity of the participants, and provide contact information if they have any questions. In addition to this, Creswell (2009) also explained that there are ethical responsibilities to the gatekeepers, which indicted that descriptions and guarantees must be provided by Qualtrics when requesting the survey be distributed.

Qualified participants who stated they had worked on an IT project in the United States in the past year were shown the informed consent information. This information was completed using a template provided by Capella University. The informed consent information included the title of the study, the researcher's e-mail and phone number, and the researcher's mentor's name and e-mail address. The informed consent included the purpose of the survey, and the fact that no information that could identify the survey respondents would be collected. There were no anticipated risks to participating further in the survey, although no study is completely risk free. This information was provided to allow participants to make the decision whether they wanted to continue with the survey or cancel out of it. A response to the question regarding informed consent was mandatory; participants answered the question to continue or closed out of the survey. If participants selected the *Yes* response, they were presented the first survey questions. If participants selected the *No* response, indicating they did not agree with the terms in the informed consent, they were presented with a thank you message and exited out of the survey.

Data Collection

Data collection began once the participant answered *Yes* to the agreement on informed consent. Participants were presented with five separate pages with 20 questions total to provide data for the variables in this study. The five pages had the following titles: project members experience; sharing knowledge; project team learning; cross-project learning; and project success. All questions were optional; however, if a participant missed answering a question a message would display informing them that an answer was left blank. This message provided awareness in case a participant missed a question accidentally and could therefore select an answer to that question before proceeding to the next page. The sixth page of the survey asked for project details and the seventh page asked for demographics. The median time to complete the survey was 5½ minutes, as reported by Qualtrics.

To ensure the data remained secure by the gatekeepers, the security statement provided by Qualtrics was reviewed. Qualtrics stores data in a specific location in a trusted data center, rather than a distributed type of storage, such as the cloud. Qualtrics servers are behind firewall systems and are scanned and patched regularly. In addition, a complete penetration test is performed annually. Qualtrics data security measures meet the requirements of many Federal Acts. In addition, the researcher has a specific login ID and complex password to access the survey and collected data from Qualtrics. The researcher downloaded the data to perform the analysis. Data in the researcher's possession is not accessible by other parties and is encrypted and backed up. Survey data in the possession of the researcher will be destroyed in seven years as required by Capella University. By following these procedures, there are no ethical issues or concerns with the sampling procedure.

Data Analysis

Data collected was transferred from Qualtrics into Microsoft Excel, and then imported into SPSS 23 and AMOS 23 for Windows for analysis. Incomplete survey responses were eliminated and 128 complete survey responses were received. The 20 questions in the survey that were used to define the variables of the research questions were Likert-type responses and obtained ordinal data.

Descriptive statistics. Descriptive statistics describe the distribution, central tendency and dispersion for each variable (Trochim & Donnelly, 2006). A table of descriptive statistics was created in SPSS. Exploratory data analysis with SPSS will allow the results to be visually inspected before running further analysis (Field, 2009). This included scatterplots, which are a graphical depiction of the correlation of two variables (Rovai, Baker, and Ponton, 2014).

For each variable, the Likert-type responses were assigned a number one through seven, or one through five depending on the number of choices for each question. For each variable the minimum, maximum, mean, standard deviation, skewness, and Kurtosis statistic were generated. The data was analyzed for normality with the Kolmogorov-Smirnov test. Prior to testing the research questions and hypotheses, bivariate correlations were conducted as a preliminary analysis to confirm the relationships between the variables were linear.

Hypothesis testing. In the process of conducting this study, the goal was to collect enough information to answer the research questions using structural equation modeling to measure the magnitude and direction of the relationship among the variables in each hypothesis statement, and to test the model presented in Figure 1 (see page 91). There were eight research questions and associated hypotheses in this study. Path analysis was used to test the hypotheses to determine the degree of the relationship between the variables. Regression weights and standardized regression weights were calculated.

Instrument

Survey Developed by Newell and Edelman (2008)

The instrument used in this study was developed by Newell and Edelman (2008). The purpose of the survey was to identify the mechanisms that can enhance project learning and cross-project learning in a dynamic environment. The company studied by Newell and Edelman (2008) was a utility company in the United Kingdom. In this research, the survey will be applied to IT projects that represent complex and dynamic environments. The survey data was tested with confirmatory factor analysis scores and Cronbach's alpha scores for each variable.

Field testing was not applicable in this research because this was not a qualitative study. Permission was obtained by Dr. Newell and Dr. Edelman to use the existing survey instrument they developed.

Pilot testing was not applicable in this research because of previous validation of the survey from previous research.

The independent variables making up knowledge transfer were operationalized as follows (see Table 3). Individual knowledge was measured with two questions with a factor score of .89 and a Cronbach's alpha of .73. Knowledge articulation was measured with two questions with a factor score of .86 and a Cronbach's alpha of .73. Knowledge codification was measured with three questions with a factor score of .72 and a Cronbach's alpha of .79.

The mediating variables making up team learning were operationalized as follows. Project learning was measured with four questions with a factor score of .77 and a Cronbach's alpha of .79. Cross-project learning was measured with two questions with a factor score of .78 and a Cronbach's alpha of .71. The dependent variable was operationalized as follows. Project success was measured with six questions with a factor score of .65 and a Cronbach's alpha of .81. The survey instrument used in this research had previously been used and tested in a prior study, which reduces measurement error. Using an existing instrument allowed for comparison for the instrument's validity and the results obtained from this study.

Table 3

Types of Data

RQ	Variables	IV/DV	Likert-type-type Scale	Data Type
1	Individual knowledge	IV	7 point	Ordinal
	Project learning	MV	7 point	Ordinal
2	Individual knowledge	IV	7 point	Ordinal
	Cross-project learning	MV	5 point	Ordinal
3	Knowledge articulation	IV	7 point	Ordinal
	Project learning	MV	7 point	Ordinal
4	Knowledge articulation	IV	7 point	Ordinal
	Cross-project learning	MV	5 point	Ordinal
5	Knowledge codification	IV	7 point	Ordinal
	Project learning	MV	7 point	Ordinal
6	Knowledge codification	IV	7 point	Ordinal
	Cross-project learning	MV	5 point	Ordinal
7	Project learning	MV	7 point	Ordinal
	Project success	DV	5 point	Ordinal
8	Cross-project learning	MV	5 point	Ordinal
	Project success	DV	5 point	Ordinal

Validity. Validity measures how well the research design answers the research questions. A threat to validity in this research is the dependency of the participants answering the survey questions truthfully, which was previously identified as a limitation. To improve the validity, the initial qualifying question of the survey was open ended. The initial question, to determine participant qualification was the following. *Which of the following work related activities have you completed in the Unites States in the past year?* Participants could select one or more of the following answers provided: negotiated a contract; worked on an IT project; provided day-to-day production support; and other. If participants did not select *worked on an information technology project,* a thank you message was presented and the survey was ended. If participants indicated they had worked on an information technology project, they were advanced to the informed consent page. It was important to determine if participants were qualified prior to the informed consent because the informed consent page explained the topic the survey was addressing. This eliminated the situation where participants indicate that they have worked on an IT project in order to take the survey, when they may not have.

A second question was added in the middle of the survey to ensure the participants were reading the questions and thinking about their answers. The question that was added stated, *This is an attention filter. Please select Red below.* If participants did not select the answer *Red*, a thank you message was displayed and the survey was closed. The attention filter question helped eliminate participants that were quickly going through the questions without reading them or thinking about their answer, just to complete the survey.

Confirmatory factor analysis was used to analyze the data to determine the validity of the instrument in this study. The analysis confirmed the presence of six unique factors, which represent the variables in this study. The six factors were individual knowledge, knowledge articulation, knowledge codification, project learning, cross-project learning, and project success. This confirms the instrument was valid; the questions consistently measured the variables represented in this study.

Reliability. According to Vogt (2007), Cronbach's alpha measures the reliability and consistency of items on a scale. In this scenario, it measures how likely the questions in the survey consistently measure the same variable. The minimal acceptable value for Cronbach's alpha is .70 (Vogt, 2007). The values measured by Newell and Edelman (2008) are all above the minimal acceptable value, indicating the survey questions were reliable and consistent in their research. For this research, Cronbach's alpha was calculated to determine the surveys reliability. The internal consistency of the variables ranged from .76 and .88. The Cronbach's alpha for all variables in this study were higher than the values presented in the original study indicating the survey was reliable. Structural equation modeling was used to show the relationships between the survey questions and the six latent variables. Standardized regression coefficients were generated and the comparative fit index (CFI) was used to access how well the data fit the model. Results of the analysis are further presented in Chapter 4.

Ethical Considerations

The *Belmont Report* was published in 1979 by the Commission for the Protection of Human Subjects of Biomedical and Behavioral

Research. The *Belmont Report* was designed to safeguard the rights of individuals involved in research, and establish cultural integrity in the research community (Edward, 2003). Three ethical principles are the foundation for the *Belmont Report*: respect for persons, beneficence, and justice.

The ethical principle of respect for persons has two parts: to respect the autonomy of individuals, and to protect individuals that may have a "diminished capacity to make autonomous decisions" (Edward, 2003, p. 20). This research did not collect any personal identifying data from survey participants, such as name, location, or company. This study followed the guidelines of Capella University's institutional review board (IRB). The survey instrument, data collection method, and statistical method were approved by the IRB prior to any data collection. Survey participants were presented a copy of the informed consent form, which required an agreement and understanding of participants before they were presented the survey questions. If individuals did not agree to the informed consent form, they were presented with a thank you message and exited from the survey. Informed consent ensures that participation in the study is voluntary and not coerced. There was no personal contact between the survey participants and the researcher.

The respect for persons is the underlying principle of the *Belmont Report* and can be thought of as doing the right thing. However, all research has some degree of risk. The second ethical principle of beneficence ensures that the research will do no harm to the participants. This means that the risks and benefits of the research must be balanced. Institutional review boards are required to determine this balance for the proposed research (Edward, 2003). In this study, there were no consequences to the participants to participate. The participants could complete the survey online at the time and location of their choosing. Participants could also quit the survey at any time. In addition, the survey questions were not mandatory, with the exception of the question regarding agreement to informed consent. If a participant felt uncomfortable in any way by answering a particular question, they could skip it.

Justice is the third ethical principle of the *Belmont Report*. Justice requires all research participants to be treated fairly; the "risks of the research can never be made to sit unfairly on any one part of the population" (Edward, 2003, p. 23). The survey was distributed by Qualtrics to their established audience. Participants were presented the qualifying question to continue the survey. Those that indicated they had worked on an IT project in the past year were presented with the informed consent form. Individuals who did not qualify for this study received a thank you message and exited the survey. All participants were treated equally. Demographic data, such as gender and age were optional questions and were included only to provide a comparison with the larger population.

Summary

This chapter included an overview of the research method, design, and procedures for conducting this study. The purpose for the study was presented with more detail, followed by the population and target sample. The research questions and hypotheses were restated along with details on the steps taken to distribute the survey, collect, and analyze the data. The survey instrument was presented with information on its reliability and validity. The data analysis results of this study are presented next in Chapter 4.

CHAPTER 4. RESULTS

Background

The purpose of this quantitative research was to test the model of knowledge transfer, team learning, and project success, which relates individual knowledge, knowledge articulation, and knowledge codification to project success controlling for projects in the IT field. This chapter will report the statistical analysis of the data collected in this study. The description of the sample, including demographics, descriptive statistics, and data screening will be presented first. Reliability and validity analysis is presented next, followed by the preliminary analysis prior to addressing each research question and hypothesis. The data were analyzed with SPSS 23 and AMOS 23 for Windows. The following provides a description of the sample.

Description of the Sample

Sample Demographics

The sample consisted of 128 survey responses from individuals who have worked on an IT project in the United States; 57% ($n = 73$) were males and 43% ($n = 55$) were females who had been working on IT projects from 2-46 years ($M = 12.61$, $SD = 8.83$) with a median of 10 years experience. Approximately 66% ($n = 84$) were 18-44 years of age and the remaining 34% (*n* = 44) were older than 44. Age is presented in Table 4. Cumulative percentage is provided because age data is ordinal.

Table 4

Age

Age	n	%	Cumulative%
18-29	18	14.1	14.1
30-44	66	51.6	65.6
45-59	36	28.1	93.8
60+	8	6.3	100.0
Total	128	100.0	

Regarding project roles, 41.4% (*n* = 53) were technical project managers, 35.9% (*n* = 46) were on a technology project team, and 7.8% (*n* = 10) were business project managers to name a few. Participant project roles are included in Table 5.

Table 5

Project Roles

Role	n	%
Business Project Manager	10	7.8
Technical Project Manager	53	41.4
Business Project Team	6	4.7
Technology Project Team	46	35.9
Business Stakeholder	1	0.8
Technology Stakeholder	3	2.3
Project Sponsor	1	0.8
Executive Project Sponsor	7	5.5
Business Analyst	1	0.8
Total	128	100.0

The largest industries/fields the projects were conducted in were technology-information services, which represented half (50.8%, $n = 65$) of the project professionals, whereas 9.4% ($n = 12$) were conducted in professional services, and 8.6% ($n = 11$) were conducted in finance and insurance. Table 6 provides the industry/field of the project.

Table 6

Industry/Field in Which Project Was Conducted

Industry/Field	n	%
Health Care	4	3.1
Non-profit	1	0.8
Technology-Information Services	65	50.8
Energy & Utilities	9	7.0
Transportation	2	1.6
Construction	4	3.1
Finance & Insurance	11	8.6
Government	5	3.9
Professional Services	12	9.4
Manufacturing	8	6.3
Education	2	1.6
Other (Distribution, software, telecommunications, etc.)	5	3.9
Total	128	100.0

Approximately 54% ($n = 69$) of the projects were considered services and 46% ($n = 59$) were considered products. The three most frequent sizes of the project teams consisted of less than 10 (26.6%, $n = 34$), 11-25 (25%, $n = 32$), and 26-50 (21.1%, $n = 27$). Size of the project team is presented in Table 7.

Table 7

Size of Project Team

Size	n	%	Cumulative%
<10	34	26.6	26.6
11-25	32	25.0	51.6
26-50	27	21.1	72.7
50-100	21	16.4	89.1
100-200	12	9.4	98.4
>200	2	1.6	100.0
Total	128	100.0	

Regarding project complexity, 41.4% ($n = 53$) indicated that the project was slightly similar to other projects, and 27.3% ($n = 35$) indicated that the project was somewhat similar to other projects. However, 16.4% ($n = 21$) of the survey respondents indicated that the project was extremely complex and had not been done before as reported in Table 8.

Table 8

Level of Project Complexity

Project Complexity	n	%
Simple-repeatable change	3	2.3
Very similar to other projects	16	12.5
Somewhat similar to other projects	35	27.3
Slightly similar to other projects	53	41.4
Extremely complex-not done before	21	16.4
Total	128	100.0

Descriptive Statistics

Scores for the variables of interest were computed by calculating the mean responses for each variable or subscale. As an example, for individual knowledge, scores ranged from 1 to 7 ($M = 5.47$, $SD = 1.38$). For project success, scores ranged from 1.17 to 5 ($M = 3.84$, $SD = 0.75$). The mean is used here because not all variables have the same number of Likert-type responses; simply adding up the scores of the variables would add bias to the results. Descriptive statistics are presented in Table 9.

Table 9

Descriptive Statistics

Variable	N Statistic	Minimum Statistic	Maximum Statistic	M Statistic	SD Statistic
Individual Knowledge	128	1.00	7.00	5.47	1.38
Knowledge Articulation	128	1.00	7.00	5.70	1.35
Knowledge Codification	128	1.00	7.00	5.10	1.38
Project Learning	128	1.00	7.00	5.24	1.22
Cross-Project Learning	128	1.50	5.00	3.65	0.83
Project Success	128	1.17	5.00	3.84	0.75

Data Screening

The data was screened for normality with the Kolmogorov-Smirnov Test of Normality to understand if the data met the assumptions of analysis. When $p < 0.05$, the distribution significantly departs from normality. As indicated in Table 10, all of the distributions for the variables of interest significantly departed from normality. However, because 128 is a large sample size, and transformations would complicate the analysis and make it difficult to interpret, no data transformations were conducted. The nature of the data was preserved.

Table 10

Kolmogorov-Smirnov Test of Normality

Variable	Statistic	df	p
Individual Knowledge	.188	128	.000
Knowledge Articulation	.194	128	.000
Knowledge Codification	.153	128	.000
Project Learning	.135	128	.000
Cross-Project Learning	.210	128	.000
Project Success	.132	128	.000

Data Analysis and Results

Reliability Analysis

The reliability of the instrument for the sample was tested with Cronbach's alpha. The internal consistency of the variables ranged from .76 for knowledge codification to .88 for project success. A value of .70 indicates minimal acceptable reliability. The high values indicate that the survey had reliable and consistent measurement of the variables. Reliability coefficients are presented in Table 11.

Table 11

Reliability Coefficients

Variable	N of Items	Cronbach's alpha
Individual Knowledge	2	.811
Knowledge Articulation	2	.844
Knowledge Codification	3	.762
Project Learning	4	.873
Cross-Project Learning	2	.796
Project Success	6	.884

Validity

Confirmatory factor analysis was conducted in order to assess the construct validity of the survey instrument. Analysis of the survey data confirmed the presence of six unique factors; F1 = individual knowledge, F2 = knowledge articulation, F3 = knowledge codification, F4 = project learning, F5 = cross-project learning, and F6 = project success. Factor score weights with the highest values are in bold and correspond to specific questions on the survey. For cross-project learning (F5) for instance, the two largest factor score weights were observed for questions Q16 and Q17. Similarly, for project learning (F4), the four largest factor score weights were observed for questions Q10, Q11, Q13, and Q14. Factor score weights are presented in Table 12.

Table 12

Factor Score Weight Matrix

Item	F5	F4	F3	F2	F1	F6
Q23	0.045	0.01	-0.038	0.013	-0.006	**0.132**
Q22	0.04	0.009	-0.034	0.012	-0.006	**0.118**
Q21	0.041	0.009	-0.035	0.012	-0.006	**0.12**
Q20	0.057	0.012	-0.048	0.017	-0.008	**0.166**
Q19	0.043	0.009	-0.036	0.013	-0.006	**0.125**
Q18	0.04	0.009	-0.034	0.012	-0.006	**0.117**
Q17	**0.208**	0.033	0.103	-0.047	0.044	0.048
Q16	**0.311**	0.049	0.154	-0.07	0.065	0.071
Q14	0.019	**0.177**	0.019	0.002	-0.013	0.006
Q13	0.029	**0.272**	0.029	0.003	-0.02	0.009
Q11	0.012	**0.114**	0.012	0.001	-0.008	0.004
Q10	0.016	**0.149**	0.016	0.001	-0.011	0.005
Q9	0.028	0.009	**0.224**	0.019	0.053	-0.011
Q8	0.021	0.007	**0.168**	0.014	0.04	-0.008
Q7	0.025	0.008	**0.199**	0.017	0.047	-0.01
Q6	-0.044	0.003	0.065	**0.438**	0.17	0.013
Q5	-0.023	0.001	0.034	**0.225**	0.088	0.007
Q4	0.016	-0.008	0.071	0.066	**0.237**	-0.002
Q3	0.025	-0.013	0.113	0.105	**0.379**	-0.004

Structural equation modeling (SEM) was used to illustrate the relationships between the survey questions and the six latent variables (F1-F6). Standardized regression coefficients ranged from .66 to .90. In structural equation modeling, several indices are used to assess the model fit. One such measure is the comparative fit index (CFI). Comparative fit index values can range from 0 to 1 with values close to 1 indicating a very good fit for the data. For the six factors, the CFI = 0.87, which is a very good fit for the data. The path diagram is presented in Figure 2. Note that e1-e19 in the diagram represents the residual error in a regression equation and double arrows represent covariance between the latent variables.

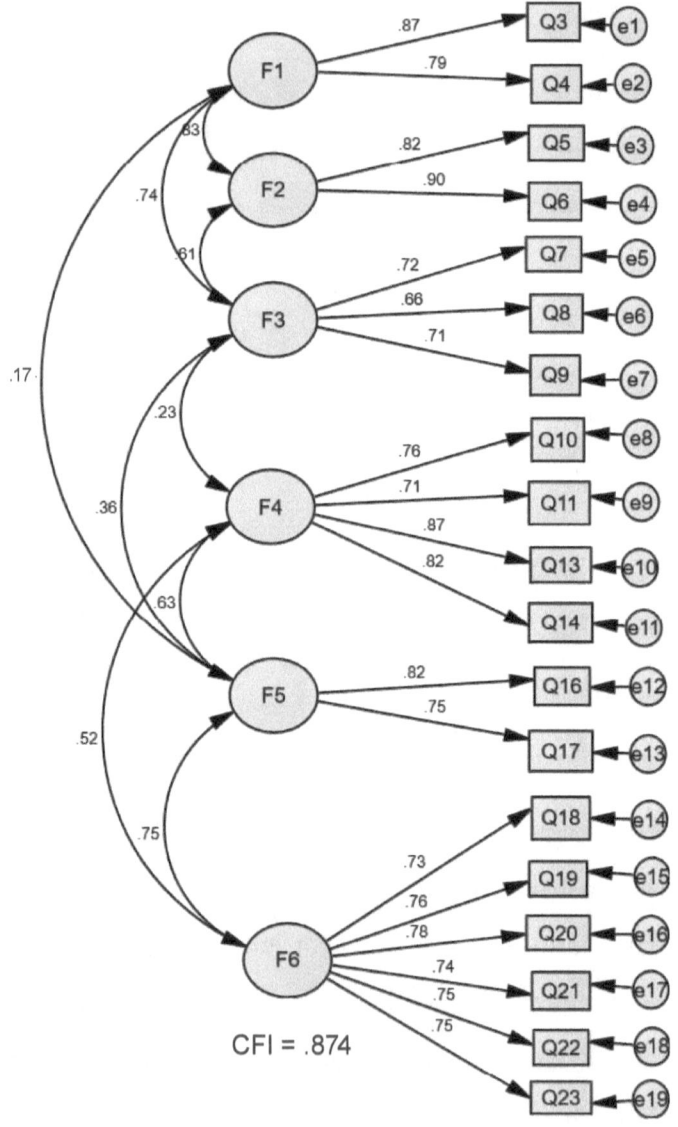

Figure 2. Confirmatory factor analysis showing the relationship between the six latent variables (F1 - F6) and the survey questions (Q#) associated with each variable. Residual error is represented by e#.

The survey questions were significantly related to their corresponding factors (latent variables) at the $p < .001$ level. In AMOS, latent variables must have a scale; this was done by setting the regression weight to 1 for first question associated with the variable. The unstandardized regression estimate, standard error, critical ratio (estimate divided by the standard error), and probability values are presented in Table 13. Critical ratios are similar to the t-statistic in that critical ratios measure the sample statistic divided by the standard error (Blunch, 2013), and the t-statistic measures the model over the error in the model (Field, 2009). A critical ratio value of greater than 1.96 will be significant at the 0.05 level.

Table 13

Significance of Factor Structure

			Estimate	S.E.	C.R.	p
Q3	<---	F1	1			
Q4	<---	F1	0.846	0.087	9.73	***
Q5	<---	F2	1			
Q6	<---	F2	1.082	0.106	10.23	***
Q7	<---	F3	1			
Q8	<---	F3	0.88	0.138	6.40	***
Q9	<---	F3	0.878	0.129	6.79	***
Q10	<---	F4	1			
Q11	<---	F4	0.956	0.121	7.89	***
Q13	<---	F4	1.278	0.131	9.78	***
Q14	<---	F4	1.25	0.135	9.24	***
Q16	<---	F5	1			
Q17	<---	F5	0.951	0.114	8.37	***
Q18	<---	F6	1			
Q19	<---	F6	1.074	0.13	8.25	***
Q20	<---	F6	0.964	0.113	8.54	***
Q21	<---	F6	1.002	0.125	8.03	***
Q22	<---	F6	1.093	0.134	8.16	***
Q23	<---	F6	0.98	0.12	8.17	***

Note. ***$p < .001$.

Preliminary Analyses

Prior to testing the research questions and hypotheses, bivariate correlations were conducted as preliminary analyses in order to confirm that the relationships between the variables of interest were linear. Inter-correlations ranged from 0.39 to 0.70 and they were all statistically significant at the $p < .001$ level. The Pearson's correlation matrix presented in Table 14 shows that all variables are inter-correlated.

Table 14

Correlation Matrix

Variable	F1	F2	F3	F4	F5	F6
Individual Knowledge (F1)	___					
Knowledge Articulation (F2)	.702***	___				
Knowledge Codification (F3)	.631***	.556***	___			
Project Learning (F4)	.610***	.555***	.614***	___		
Cross-Project Learning (F5)	.529***	.393***	.607***	.580***	___	
Project Success (F6)	.549***	.470***	.506***	.509***	.679***	___

Note. $p < .001$, two-tailed; $N = 128$.

Hypothesis Testing

The omnibus research question was as follows: *How does the model of knowledge transfer, team learning, and project success explain the relationship between project success and individual knowledge, knowledge articulation, and knowledge codification, controlling for the effects of information technology projects?* There were eight associated research questions and hypotheses. Path analysis with the maximum likelihood (ML) estimator was used to test the hypotheses and to determine the degree of the relationships between the variables. The overall model was built with each successive question. The research questions and hypotheses were as follows:

RES Q1: What is the relationship, if any, between individual knowledge and project learning?

$H1_0$: There is no significant relationship between individual knowledge and project learning.

$H1_A$: There is a significant relationship between individual knowledge and project learning.

RES Q2: What is the relationship, if any, between individual knowledge and cross-project learning?

$H2_0$: There is no significant relationship between individual knowledge and cross-project learning.

$H2_A$: There is a significant relationship between individual knowledge and cross-project learning.

RES Q3: What is the relationship, if any, between knowledge articulation and project learning?

$H3_0$: There is no significant relationship between knowledge articulation and project learning.

$H3_A$: There is a significant relationship between knowledge articulation and project learning.

RES Q4: What is the relationship, if any, between knowledge articulation and cross-project learning?

$H4_0$: There is no significant relationship between knowledge articulation and cross-project learning.

$H4_A$: There is a significant relationship between knowledge articulation and cross-project learning.

RES Q5: What is the relationship, if any, between knowledge codification and project learning?

$H5_0$: There is no significant relationship between knowledge codification and project learning.

$H5_A$: There is a significant relationship between knowledge codification and project learning.

RES Q6: What is the relationship, if any, between knowledge codification and cross-project learning?

$H6_0$: There is no significant relationship between knowledge codification and cross-project learning.

$H6_A$: There is a significant relationship between knowledge codification and cross-project learning.

RES Q7: What is the relationship, if any, between project learning and project success?

$H7_0$: There is no significant relationship between project learning and project success.

$H7_A$: There is a significant relationship between project learning and project success.

RES Q8: What is the relationship, if any, between cross-project learning and project success?

$H8_0$: There is no significant relationship between cross-project learning and project success.

$H8_A$: There is a significant relationship between cross-project learning and project success.

Research Question 1/Hypothesis 1

What is the relationship, if any, between individual knowledge and project learning? There was a significant, positive relationship between individual knowledge and project learning, (*Estimate* = .54, *C.R.* = 8.68, $p < .001$); $R^2 = .37$. Critical ratio is greater than 1.96 and the *p* value is greater than 0.05 showing that the relationship is statistically significant and not based on chance. R^2 is the coefficient of determination that represents how much variance is explained by the analysis. For this question, 37% of project learning can be explained by individual knowledge. The unstandardized regression coefficient is presented in Table 15.

Table 15

Regression Weights for Research Question 1/Hypothesis 1

			Estimate	S.E.	C.R.	p
Project Learning	<---	Individual Knowledge	.537	.062	8.68	***

Note. *** $p < .001$.

As individual knowledge increases by one unit, project learning increases by 0.54 units. Using the standardized regression coefficient interpretation, we can say that as individual knowledge increases by 1 unit, project learning increases by 0.61 standard deviations. (Recall that these two variables are inter-correlated from Table 14.) The path diagram is presented in Figure 3.

$H1_0$ stated that there is no significant relationship between individual knowledge and project learning. There was a significant, positive relationship between individual knowledge and project learning, (*Estimate* = .54, *C.R.* = 8.68, $p < .001$). Therefore, the null hypothesis was rejected.

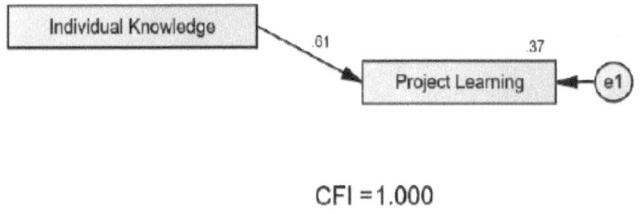

CFI = 1.000

Figure 3. Path diagram for Research Question 1/Hypothesis 1 showing the relationship between individual knowledge and project learning. Residual error is shown as e1.

Research Question 2/Hypothesis 2

What is the relationship, if any, between individual knowledge and cross-project learning? There was a significant, positive relationship between individual knowledge and cross-project learning, (*Estimate* = .32, *C.R.* = 7.03, *p* < .001); R^2 = .28. Unstandardized regression coefficients are presented in Table 16.

Table 16

Regression Weights for Research Question 2/Hypothesis 2

			Estimate	S.E.	C.R.	p
Project Learning	<---	Individual Knowledge	.537	.062	8.68	***
Cross-Project Learning	<---	Individual Knowledge	.319	.045	7.03	***

Note. *** *p* < .001.

As individual knowledge increases by one unit, cross-project learning increases by 0.32 units. The path diagram is presented in Figure 4.

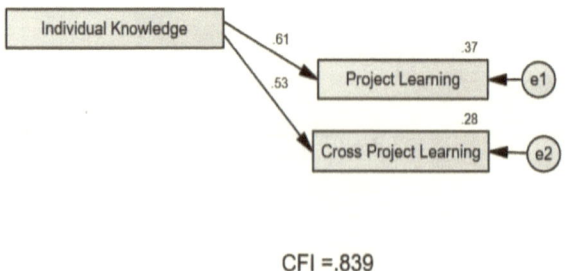

CFI =.839

Figure 4. Path diagram for Research Question 2/Hypothesis 2 showing the relationship between individual knowledge with project learning and cross-project learning. Residual error is shown as e#.

H2$_0$ stated that there is no significant relationship between individual knowledge and cross-project learning. There was a significant, positive relationship between individual knowledge and cross-project learning, (*Estimate* = .32, *C.R.* = 7.03, $p < .001$). Therefore, the null hypothesis was rejected.

Research Question 3/Hypothesis 3

What is the relationship, if any, between knowledge articulation and project learning? There was a significant, positive relationship between knowledge articulation and project learning, (*Estimate* = .23, *C.R.* = 2.61, $p = .009$); $R^2 = .40$. A probability (*p*) value of less than 0.05 is statistically significant. Unstandardized regression coefficients are presented in Table 17.

Table 17

Regression Weights for Research Question 3/Hypothesis 3

			Estimate	S.E.	C.R.	p
Project Learning	<---	Individual Knowledge	.382	.085	4.52	***
Cross-Project Learning	<---	Individual Knowledge	.319	.045	7.03	***
Project Learning	<---	Knowledge Articulation	.227	.087	2.61	.009

Note. ***$p < .001$.

As project learning increases by one unit, knowledge articulation increases by 0.23 units. The path diagram is presented in Figure 5.

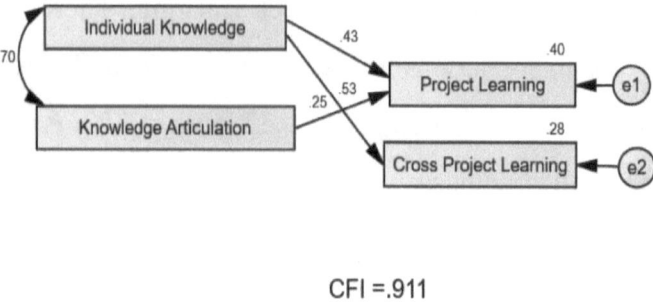

Figure 5. Path diagram for Research Question 3/Hypothesis 3 showing the relationship of individual knowledge with project learning and cross-project learning when the relationship between knowledge articulation with project learning is included. Residual error is noted as e#.

Since the model is being built with each successive question analysis, the inter-correlations in the diagrams are changing. The variables were inter-correlated in bivariate analysis as previously shown in Table 14; however, multivariate analysis produces different results. This is not a contradiction to the preliminary analysis; indeed, the bivariate correlation between individual knowledge and knowledge articulation is now shown. As the model is built with each successive question, the attention should be given to the last row of the regression weight tables.

$H3_0$ stated that there is no significant relationship between knowledge articulation and project learning. There was a significant, positive relationship between knowledge articulation and project learning, (*Estimate* = .23, *C.R.* = 2.61, $p < .009$). Therefore, the null hypothesis was rejected.

Research Question 4/Hypothesis 4

What is the relationship, if any, between knowledge articulation and cross-project learning? There was no significant relationship between knowledge articulation and cross-project learning, (*Estimate* = .03, C.R. = 0.40, $p < .689$). Unstandardized regression coefficients are presented in Table 18.

Table 18

Regression Weights for Research Question 4/Hypothesis 4

		Estimate	S.E.	C.R.	p
Project Learning <--- Individual Knowledge		.382	.085	4.52	***
Cross-Project Learning <--- Individual Knowledge		.301	.064	4.73	***
Project Learning <--- Knowledge Articulation		.227	.087	2.61	.009
Cross-Project Learning <--- Knowledge Articulation		.026	.065	0.40	.689

Note. ***$p < .001$.

The unstandardized (.03) and standardized (.04) regression coefficients for knowledge articulation regressing on cross-project learning are minimal. The path diagram is presented in Figure 6.

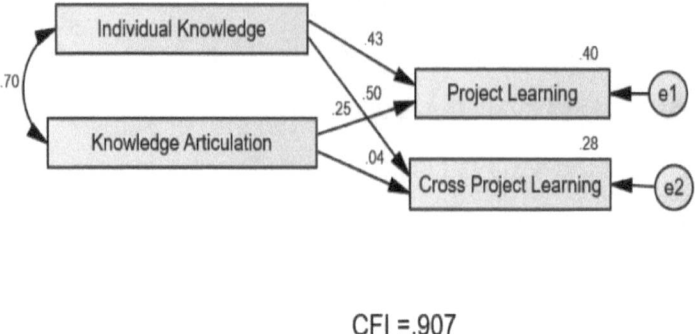

Figure 6. Path diagram for Research Question 4/Hypothesis 4 showing the combined relationship of both individual knowledge and knowledge articulation with project learning and cross-project learning. Residual error is noted as e#.

H4$_0$ stated that there is no significant relationship between knowledge articulation and cross-project learning. There was no significant relationship between knowledge articulation and cross-project learning, (*Estimate* = .03, *C.R.* = 0.40, *p* < .689). The critical ratio was not above 1.96 and the *p* value was above 0.05 indicating the relationship was not statistically significant. Therefore, the null hypothesis was not rejected.

Research Question 5/Hypothesis 5

What is the relationship, if any, between knowledge codification and project learning? There was a significant, positive relationship between knowledge codification and project learning (*Estimate* = .31, *C.R.* = 4.11, *p* < .001); R^2 = .47. Unstandardized regression coefficients are presented in Table 19.

As project learning increases by one unit, knowledge codification increases by .31 units. The path diagram is presented in Figure 7. Here the inter-correlations between the three independent variables are shown; if they were not considered the fit for the model (CFI) would be lower.

Table 19

Regression Weights for Research Question 5/Hypothesis 5

			Estimate	S.E.	C.R.	p
Project Learning	<---	Individual Knowledge	.237	.087	2.72	.006
Cross-Project Learning	<---	Individual Knowledge	.301	.064	4.73	***
Project Learning	<---	Knowledge Articulation	.156	.083	1.87	.061
Cross-Project Learning	<---	Knowledge Articulation	.026	.065	0.40	.689
Project Learning	<---	Knowledge Codification	.308	.075	4.11	***

Note. ***p < .001.

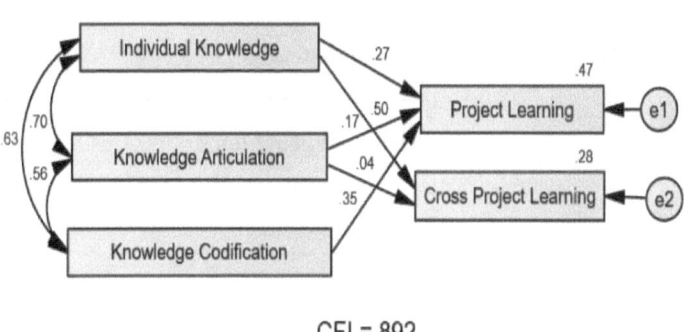

Figure 7. Path diagram for Research Question 5/Hypothesis 5 showing the relationships between individual knowledge and knowledge articulation with both project learning and cross-project learning, after the relationship of knowledge codification with project learning has been considered. Residual error is noted as e#.

H5$_0$ stated that there is no significant relationship between knowledge codification and project learning. There was a significant, positive relationship between knowledge codification and project learning (*Estimate* = .31, *C.R.* = 4.11, *p* < .001). Therefore, the null hypothesis was rejected.

Research Question 6/Hypothesis 6

What is the relationship, if any, between knowledge codification and cross-project learning? There was a significant, positive relationship between knowledge codification and cross-project learning, (*Estimate* = .28, *C.R.* = 5.16, *p* < .001); R^2 = .41. Unstandardized regression coefficients are presented in Table 20.

Table 20

Regression Weights for Research Question 6/Hypothesis 6

			Estimate	S.E.	C.R.	p
Project Learning	<---	Individual Knowledge	.237	.087	2.72	.006
Cross-Project Learning	<---	Individual Knowledge	.168	.063	2.66	.008
Project Learning	<---	Knowledge Articulation	.156	.083	1.87	.061
Cross-Project Learning	<---	Knowledge Articulation	-.038	.061	-0.63	.532
Project Learning	<---	Knowledge Codification	.308	.075	4.11	***
Cross-Project Learning	<---	Knowledge Codification	.281	.054	5.16	***

Note. ***p < .001.

As cross-project learning increases by one unit, knowledge codification increases by 0.28 units. The path diagram is presented in Figure 8.

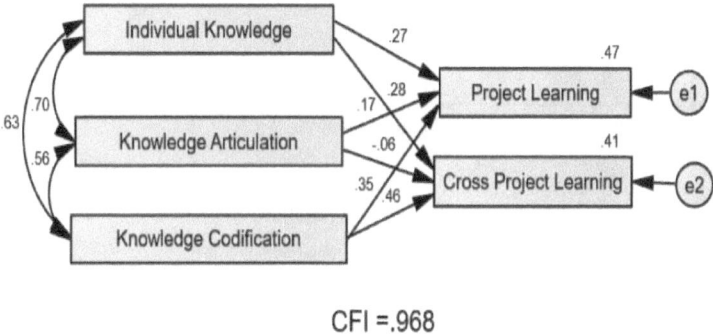

Figure 8. Path diagram for Research Question 6/Hypothesis 6 showing the combined relationship of the construct of knowledge transfer comprised of individual knowledge, knowledge articulation, and knowledge codification, with both project learning and cross-project learning. Residual error is depicted as e#.

$H6_0$ stated that there is no significant relationship between knowledge codification and cross-project learning. There was a significant, positive relationship between knowledge codification and project learning (*Estimate* = .28, *C.R.* = 5.16, $p < .001$). Therefore, the null hypothesis was rejected.

Research Question 7/Hypothesis 7

What is the relationship, if any, between project learning and project success? There was a significant, positive relationship between project learning and project success, (*Estimate* = .31, *C.R.* = 6.67, $p < .002$); $R^2 = .26$. Unstandardized regression coefficients are presented in Table 21.

Table 21

Regression Weights for Research Question 7/Hypothesis 7

		Estimate	S.E.	C.R.	p
Project Learning	<--- Individual Knowledge	.237	.087	2.72	.006
Project Learning	<--- Knowledge Articulation	.156	.083	1.87	.061
Project Learning	<--- Knowledge Codification	.308	.075	4.11	***
Cross-Project Learning	<--- Individual Knowledge	.168	.063	2.66	.008
Cross-Project Learning	<--- Knowledge Articulation	-.038	.061	-0.63	.532
Cross-Project Learning	<--- Knowledge Codification	.281	.054	5.16	***
Project Success	<--- Project Learning	.314	.047	6.67	***

Note. ***p < .001.

As project success increases by one unit, project learning increases by 0.31 units. The path diagram is presented in Figure 9. $H7_0$ stated that there is no significant relationship between project learning and project success. There was a significant, positive relationship between project learning and project success, (*Estimate* = .31, *C.R.* = 6.67, *p* < .001). Therefore, the null hypothesis was rejected.

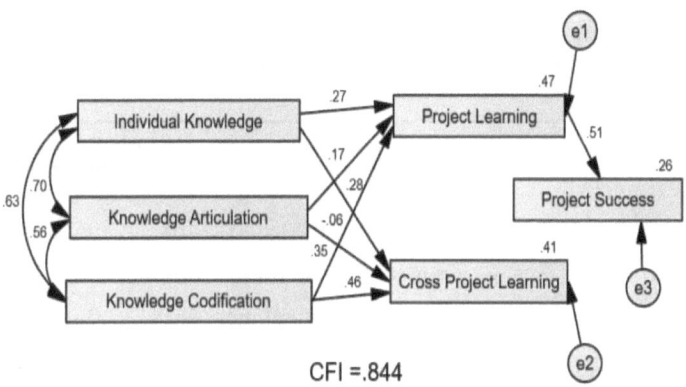

Figure 9. Path diagram for Research Question 7/Hypothesis 7 showing the combined relationship of the three components of knowledge transfer with team learning. This diagram includes both project learning and cross-project learning, after the effect of project learning with project success had been considered. Residual errors are illustrated as e#.

Research Question 8/Hypothesis 8

What is the relationship, if any, between cross-project learning and project success? There was a significant, positive relationship between cross-project learning and project success, (*Estimate* = .52, *C.R.* = 8.21, $p < .001$); $R^2 = .46$. Unstandardized regression coefficients are presented in Table 22. As project success increases by one unit, cross-project learning increases by 0.52 units. The path diagram is presented in Figure 10.

Table 22

Regression Weights for Research Question 8/Hypothesis 8

			Estimate	S.E.	C.R.	p
Project Learning	<---	Individual Knowledge	.237	.087	2.72	.006
Cross-Project Learning	<---	Individual Knowledge	.168	.063	2.66	.008
Project Learning	<---	Knowledge Articulation	.156	.083	1.87	.061
Cross-Project Learning	<---	Knowledge Articulation	-.038	.061	-0.63	.532
Project Learning	<---	Knowledge Codification	.308	.075	4.11	***
Cross-Project Learning	<---	Knowledge Codification	.281	.054	5.16	***
Project Success	<---	Project Learning	.107	.044	2.47	.014
Project Success	<---	Cross-Project Learning	.522	.064	8.21	***

Note. ***$p < .001$.

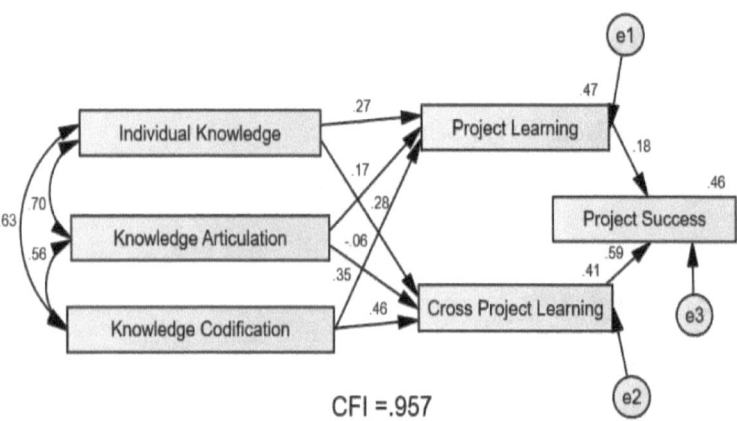

Figure 10. Path diagram for Research Question 8/Hypothesis 8 showing the combined relationship of the three variables of knowledge transfer with the two variables of team learning after the relationships between team learning and project success have been considered. Residual errors are indicated by e#.

$H8_0$ stated that there is no significant relationship between cross-project learning and project success. There was a significant, positive relationship between cross-project learning and project success, (*Estimate = .52, C.R. = 8.21, p < .001*). Therefore, the null hypothesis was rejected. Table 23 provides a summary of the hypotheses tested and the outcomes.

Table 23

Summary of Hypotheses Tested and Outcomes

Hypothesis	CFI	Significance	Outcome
$H1_0$: There is no significant relationship between individual knowledge and project learning.	1.00	$p < .001$	Null Rejected.
$H2_0$: There is no significant relationship between individual knowledge and cross-project learning.	.84	$p < .001$	Null Rejected.
$H3_0$: There is no significant relationship between knowledge articulation and project learning.	.91	$p = .009$	Null Rejected.
$H4_0$: There is no significant relationship between knowledge articulation and cross-project learning.	.91	$p = .689$	Null Not Rejected.
$H5_0$: There is no significant relationship between knowledge codification and project learning.	.89	$p < .001$	Null Rejected.
$H6_0$: There is no significant relationship between knowledge codification and cross-project learning.	.97	$p < .001$	Null Rejected.
$H7_0$: There is no significant relationship between project learning and project success.	.84	$p < .001$	Null Rejected.
$H8_0$: There is no significant relationship between cross-project learning and project success.	.96	$p < .001$	Null Rejected.

Summary

The model of knowledge transfer, team learning, and project success explains the relationship between project success and individual knowledge, knowledge articulation, and knowledge codification, controlling for the effects of information technology projects. The final model can explain 46% of the variance in project success.

- There was a significant, positive relationship between individual knowledge and project learning.
- There was a significant, positive relationship between individual knowledge and cross-project learning.
- There was a significant, positive relationship between knowledge articulation and project learning.
- There was no significant relationship between knowledge articulation and cross-project learning.
- There was a significant, positive relationship between knowledge codification and project learning.
- There was a significant, positive relationship between knowledge codification and cross-project learning.
- There was a significant, positive relationship between project learning and project success.
- There was a significant, positive relationship between cross-project learning and project success.

Implications are discussed in Chapter 5.

CHAPTER 5. DISCUSSION, IMPLICATIONS, RECOMMENDATIONS

Introduction

This final chapter will assess this study regarding how well the findings addressed the research questions, and make recommendations for improving this study and for future research. This chapter will begin with an overall summary of the results and describe how they related to Chapter 1 and Chapter 2. Following the summary, the inward focus on this research is presented in the Discussions of the Results section, and an outward focus in the Conclusions Based on the Results section. Limitations of this study is offered next, providing transparency on how this study could be improved. The implications of this research to the project management field and to project managers is addressed before the chapter concludes with a concise summary of this dissertation.

Summary of the Results

The general problem this study aimed to address is the high failure rate of IT projects. Companies spend time and money on IT projects; therefore, when IT projects fail, companies experience sunk costs and missed opportunities. Since the IT project management field is growing and spanning across industries, this is a problem worth addressing. This study focused on knowledge management in the IT project management field because projects are mainly knowledge work, and project success depends on the level of knowledge transfer (Gemino et al., 2008; Landaeta, 2008).

A review of the literature beginning with the theory of organizational learning was presented in Chapter 2. Organizational learning theory expanded into different approaches to knowledge management. However, there is no overarching theory of knowledge management, nor of project management. A gap in the literature was found relating concepts of organizational learning to practical project management (Jugdev & Mathur, 2013; Paramkusham & Gordon, 2013; Reich, 2007), specifically relating to project team learning and cross-project team learning (Mueller, 2015; Swan et al., 2010).

Dynamic capabilities are based on the theory of organizational learning and are an expansion of the concepts in the knowledge creation theory (Teece et al., 1997; Zollo & Winter, 2002). The term dynamic capabilities was introduced in the literature in 1997 as an ability (Teece et al., 1997) and expanded into a process in 2002 (Zollo & Winter, 2002). Processes are easier to measure than abilities and one of the few "advanced survey instruments" to measure dynamic capabilities was created by Newell and Edelman in 2008 (Eriksson, 2013, p. 14). The survey instrument developed by Newell and Edelman (2008) was used to measure knowledge transfer and team learning in relation to project success. This study attempted to build on that research, and gain new insights to dynamic capabilities in the IT project field. This study attempted to contribute to the field of project management and knowledge management by increasing the understanding of how knowledge management in IT projects influenced project success.

The omnibus research question was as follows: *How does the model of knowledge transfer, team learning, and project success explain the relationship between project success and individual knowledge, knowledge articulation, and knowledge codification, controlling for the effects of information technology projects?* There were eight associated research questions and hypotheses. Path analysis was used to test the hypotheses and to determine the degree of the relationships between the variables. Structural equation modeling was the method used in this research to test how well the data fit the model presented in Figure 1. Structural equation modeling has been used in several other studies regarding knowledge management and project management (Lichtenthaler, 2009; Mahaney & Lederer, 2011; Zheng et al., 2011).

All variables used in this study were found to be inter-correlated. Seven null hypothesis were rejected since the relationship between the variables was found to be significant and positive. One null hypothesis statement ($H4_0$) was not rejected; there was no significant relationship found between knowledge articulation and cross-project learning. The complete path diagram was presented in Figure 9 and reported that 46% of the variance in project success can be explained by the model. Since the 2015 CHAOS Report published by The Standish Group stated a success rate of IT projects from 27% to 31% during the last five years (Hastie & Wojewoda, 2015), the 46% impact to project success shown in this research should benefit the project management field.

Discussion of the Results

The sample size of 128 was adequate for structural equation modeling analysis due to the simplicity of the model and the number of variables. However, a larger sample size may have increased the normal distribution of the data. The project roles indicated by the survey participates indicated 90% were either the business or technical project manager. The project manager should have excellent exposure to knowledge transfer activities so this high percentage is fortunate. The other project roles indicated on the survey included a technical stakeholder or sponsor; those individuals may not have as much exposure to knowledge transfer activities.

The data obtained was not normally distributed. However, a larger sample size may still not produce normally distributed data. One can speculate that when people look back on a project they have worked on they would be most likely to think about their successes and accomplishments rather than failures. The survey is limited as it expects participants to be unbiased. Moreover, the questions do not measure actual activities that have happened during a project but only the participant's perceptions of those activities.

The survey instrument was exceptional (Eriksson, 2013); the reliability coefficients were all above .70 for each variable. The lowest reliability coefficient was 0.762 for knowledge codification. The confirmatory factor analysis tested the validity of the instrument and showed that the survey questions measure the defined variables.

The preliminary analysis discovered that all the variables were inter-correlated. The three highest correlations involved cross-project learning and individual knowledge. It is interesting that cross-project learning had a higher correlation to project success than project learning, indicating a need for knowledge outside the project team. The high correlation of individual knowledge-to-knowledge articulation and to knowledge codification is understandable and expected. Knowledge must be obtained and internalized by an individual before it can be shared by articulation or codification.

Knowledge articulation was not highly correlated. Knowledge articulation had the lowest correlation to cross-project learning, indicating that individual knowledge and knowledge codification are more important for cross-project learning. Knowledge articulation also had a low correlation for project success. There could be many reasons for this, one may be that the project team is not co-located, making casual discussions difficult. The survey did not ask if the project team was co-located or virtual so this is only a speculation. Another contributing factor may be that individuals do not readily perceive informal project conversations as a knowledge articulation activity. If so, the occurrence of knowledge articulation may be more important that reported by the survey participants.

Knowledge codification and project learning were correlated to project success but not as high as individual knowledge or cross-project learning. The high correlation of cross-project learning to project success is surprising. The high correlation of individual knowledge to project success indicates that it is important for organizations to structure their project team and assign experienced individuals to the appropriate projects to obtain project goals.

The omnibus research question was as follows: *How does the model of knowledge transfer, team learning, and project success explain the relationship between project success and individual knowledge, knowledge articulation, and knowledge codification, controlling for the effects of information technology projects?* There were eight associated research questions and hypotheses. All null hypotheses were rejected except $H4_0$. The path analysis revealed a significant and positive relationship between knowledge transfer and project learning to project success, with the exception of the relationship between knowledge articulation and cross-project learning.

Individual knowledge could explain 37% of project learning and 28% of cross-project learning. The combination of individual knowledge and knowledge articulation could explain 40% of project learning. However, knowledge articulation did not increase the 28% of cross-project learning that was explained by individual knowledge.

This was surprising, as the networking in organizations has been deemed beneficial to allow individuals to have discussions across projects. It can be speculated that these discussions help to identify individuals with beneficial knowledge and skills to add to the project team, but the discussions alone are not significant. Other possible considerations for the low effect of knowledge articulation may be due to personality conflicts, lack of communication abilities, lack of time availability, experience gaps, and lack of willingness to share knowledge.

The construct of knowledge transfer, which includes individual knowledge, knowledge articulation, and knowledge codification, accounted for 47% of project learning. Knowledge transfer was expected to have a higher contribution to project learning but the results show other factors exist that contribute to project learning. One explanation could be adjustments to activities based on experience that is not tied to one individual, and is not therefore included in individual knowledge. The construct of knowledge transfer accounted for 41% of cross-project learning, also indicating there are other factors that contribute to cross-project learning.

The construct of team learning, which includes project learning and cross-project learning accounted for 46% of project success. While this percentage is lower than expected, it is still useful as project success rates have been reported to be 27% to 31% (Hastie & Wojewoda, 2015). There are other factors that impacted project success. The survey asked six questions for project success. The questions considered creating innovative solutions, meeting time scales, meeting project objectives, adding value, staying within budget, and satisfying the client. A possible explanation could be that 83.6% of the projects considered were simple or similar to other projects, in which case team learning may have had a more significant impact to the success of past projects and have already been in place during the time of the project considered.

Ignoring non-normality and using the maximum likelihood (ML) estimator in AMOS software is an option; an alternative estimator of weighted least squares (WLS) was not considered because it would require the sample to be very large ($N>1000$) (Andreassen, Lorentzen, & Olsson, 2006). Andreassen, Lorentzen, and Olsson (2006) analyzed the impact of non-normality on structural equation modeling analysis and found that using maximum likelihood results in poor fit; transforming the data to normal scores did not increase the fit in their study, but did change the estimates. The comparative fit index for the final model in this study was .957 indicating the data fit the model. A comparative fit index higher than 0.95 is desired (Blunch, 2013; Kline, 2011).

Conclusions Based on the Results

Comparison of the Findings With the Theoretical Framework and Previous Literature

This research extends previous research on organizational learning, dynamic capabilities, and project management, by conceptualizing dynamic capabilities as knowledge transfer activities in project management. Furthermore, this research theorizes that knowledge transfer activities affect learning capabilities within and across projects, leading to project success. The data supported the model indicating that knowledge transfer activities impacted both project learning and cross-project learning which contributes to project success in the IT field. This information is useful for practitioners to understand that knowledge transfer methods can benefit project success and time and effort used for knowledge transfer should be valued. The results of this study also contribute to the scholarly research by filling a gap in the existing literature that applied dynamic capabilities in IT project management.

Newell and Edelman (2008) previously validated the survey used in this study. A comparison of the Cronbach's alpha obtained on the variables of the survey is listed in Table 24. The high Cronbach's alpha indicates the high validity of the survey instrument.

In contrast, there were differences in the individual correlations of the variables. The findings from this study found that all the variables were inter-correlated (Table 14). The previous study did not find a significant inter-correlation between three pairs of variables: individual knowledge and knowledge articulation; individual knowledge and project success; and knowledge articulation and project success (Newell & Edelman, 2008).

Table 24

Comparison of Cronbach's Alpha

Variable	Cronbach's alpha	
	N&E (2008)	This Study
Individual Knowledge	.73	.81
Knowledge Articulation	.73	.84
Knowledge Codification	.79	.76
Project Learning	.79	.87
Cross-project Learning	.71	.79
Project Success	.81	.88

[a] Individual knowledge was referred to as experience accumulation by Newell and Edelman (2008).

It is interesting that Newell and Edelman (2008) found a significant positive relationship between individual knowledge and cross-project learning, but did not find a significant positive relationship between individual knowledge and project learning. As reported in Chapter 4, this study found that 37% of project learning and 28% of cross-project learning could be explained by individual knowledge.

It was expected that individual knowledge would contribute significantly to both project learning and cross-project learning. Having the right people on the team has been cited in the literature as an important consideration to success. Ahern et al. (2015) stated that 90% of project failures are attributed to people. However, the Delphi study conducted by Hadaya et al. (2012) reported that neither scholars nor professionals rated the project manager, the project team, or the knowledge they possess as a top priority for project success. In their study of IT firms, Breznik and Lahovnik (2014) found that managerial and human resource capabilities are important for the firm's success, but are more complex and difficult to develop.

Neither this study, nor the study by Newell and Edelman (2008) obtained a significant positive relationship between knowledge articulation and cross-project learning. Furthermore, Newell and Edelman (2008) did not find a significant positive relationship between knowledge articulation and project learning. In this study, knowledge articulation was found to be significant and positively related to project learning but the relationship was not as strong ($p = .009$). When combined in the path diagram with individual knowledge, knowledge articulation only raised the percentage effect of project success from 37% to 40%.

During formal project reviews and lessons learned sessions, knowledge is articulated and codified. Both studies revealed significant and positive relationships between knowledge codification and project learning and between knowledge codification and cross-project learning. The activity of codifying knowledge creates knowledge by requiring one to consider what was done and why it was done (Breznik & Lahovnik, 2014; Zollo & Winter, 2002). Therefore, the significant positive relationship found in both studies was expected and in agreement with the research done by Cacciatori, Tamoschus, and Grabher (2011) that found knowledge codification is a means to transfer learning both within and across project teams. Eisenhardt and Martin (2000) agreed that moderately dynamic environments had a strong reliance on codification. The case study by

Hall et al. (2012) also found knowledge codification to be associated with cross-project learning.

Both studies revealed significant and positive relationships between project learning and project success, and cross-project learning and project success. These results were also expected, however, both alternatives could be found in the literature. On the one hand, trust, joint problem solving, and commitment were found to be positively and significantly related to knowledge-based dynamic capabilities (Zheng et al., 2011). Trust, joint problem solving, and commitment are expectations found within a project team. Although Zheng et al. (2011) measured knowledge-based dynamic capabilities as knowledge acquisition, generation, and combination; the results of this study are consistent with their findings.

On the other hand, Landaeta (2008) did not find cross-project learning to be significant to project success. Landaeta (2008) speculated that project learning and cross-project learning to be beneficial but limited in that too much time focusing on learning activities takes time away from working on project goals. Rungi (2014) also found that project teamwork had a negative effect on project schedule, scope, and cost.

Interpretation of the Findings

Individual learning during the project, or having experience prior to the project, could explain the influence of individual knowledge on project learning and cross-project learning. Newell and Edelman

(2008) showed that individuals did learn from their project experiences, even though they did not find support for the relationship between individual learning and project success. A suggestion was that the knowledge of the team members is valuable and should be taken into account when selecting the project team. Individual knowledge is the least expensive way to increase knowledge, especially when tasks are similar and frequent (Zollo & Winter, 2002). However, many times individuals are assigned simply based on supply and demand (Newell & Edelman, 2008). If the IT field includes more requirements for technology skills and experience that may account for the relationship between individual knowledge and both project learning and cross-project learning in this study.

The finding that knowledge articulation did not contribute to cross-project learning was unexpected, even though it was consistent with Newell and Edelman (2008). The lack of knowledge articulation may contribute to these findings; only a small amount of articulated knowledge is actually articulated (Zollo & Winter, 2002). The interview data obtained by Newell and Edelman (2008) reported that when knowledge is articulated it was not recognized as a specific activity but rather as business as usual. Knowledge articulation, may be insufficient by itself, but is required before knowledge is codified (Newell & Edelman, 2008).

Another possible consideration for this result is a trust factor. Individuals may likely consider codified knowledge more trustworthy than articulated knowledge, especially if they are unfamiliar with the person articulating. This would help explain why knowledge articulation was positive and significantly related to project team learning and not cross-project team learning. Mahaney and Lederer (2011) suggested that individuals might withhold information and act in their own interest. Withholding information may occur in the IT field where individuals are less likely to articulate knowledge across projects when they do not share project goals. If IT projects require more technical skills and experience, as suggested above, individuals in the IT field may be reluctant to share knowledge in an attempt to maintain their self-value to the organization.

Another factor to consider is the lack of availability of time to discuss problems and solutions that could be beneficial to other projects (Newell & Edelman, 2008). On the other hand, if too much time was spent, as Landaeta (2008) suggested, it may have a negative effect on the project success. If the time constraint speculated by Landaeta (2008) is true, that could explain why the research did not find a higher contributing percentage of both project learning and cross-project learning on project success.

Limitations

This study may have shed more light regarding knowledge articulation if a mixed method approach was used, such as by adding interview questions. The results that knowledge articulation did not contribute to cross-project learning does not explain if knowledge articulation occurred, or if it occurred but was not useful. Dinur

(2011) found that knowledge transfer requires a pull from the knowledge recipient. If this is true, the results could be interpreted that cross-project team members did not seek additional knowledge. This study was limited to a quantitative survey method due to time and cost considerations.

Other factors not addressed in this study include the nature of the environment. Since organizational learning is situation based (Holsapple, 2003; P. Jackson & Klobas, 2008; Jugdev & Mathur, 2013; Morris, 2013), it stands to reason that the results of knowledge transfer should be dependent on the environment. This study assumed IT projects across industries were equal. However, the degree to which an organization has a structured project management methodology would seem likely to produce different results. The IT capabilities of the organization were not considered. One would expect organizations that have a large IT infrastructure to have more experienced and skilled individuals in IT related activities. Furthermore, it is unclear whether the project team being virtual or co-located would change the results.

Implications for Practice

The high failure rate of IT projects in organizations results in sunk costs and missed opportunities. Project managers and project stakeholders should make efforts to increase knowledge transfer in projects to help reduce the project failure rate. This study presented results that show that individual knowledge, knowledge articulation, and knowledge codification all contribute, in varying degrees, to team learning and thus to project success.

First, it is important to point out that project success requirements should not be limited to time, schedule, and cost (Morris, 2013; Pinto & Slevin, 1988a). Quality of the final product or service is an important criterion for project success, which may not be fully realized immediately after the project closes. Individual knowledge was highly correlated to knowledge articulation, and had a significant impact to both project learning and cross-project learning. Therefore, individual skills and experience should be considered when selecting people for the project team to increase knowledge transfer. Organizations should invest in individual training and development to increase the individual knowledge available to the organization. Nevertheless, many organizations do not have a defined training plan or a career path for project managers (PMI, 2016).

Knowledge articulation is important for project learning but the findings did not show a contribution to cross-project learning. Knowledge articulation increases with the development of network ties (Liu et al., 2010). Knowledge articulation is an important transfer method, especially for tacit knowledge (Dinur, 2011). For that reason, time or events established for the project team to develop network ties within the project team, and across project teams, would be beneficial. Knowledge articulation is needed for knowledge codification.

Additionally, knowledge codification significantly influenced both project learning and cross-project learning. This adds support for the development of a structured project management methodology that requires project managers to complete routine documents at each phase of the project and at project closure. Often project closure documents are not completed or completed poorly due to the project team disbanding and team members being assigned to the next project (Newell et al., 2006). Knowledge codification is useful for cross-project learning and time and effort should be taken to complete knowledge codification at the end of the project.

Project learning and cross-project learning both contribute to project success. Project managers and project stakeholders should keep this in mind when projects are experiencing problems or issues. Knowledge from past projects and similar projects should be sought after to reduce the chances of repeating mistakes or reinventing the wheel. Organizations may see value in creating a forum where this knowledge could be found and shared.

Finally, the findings of this study give support to the suggestion that knowledge management be added to the PMBOK. The *Project Management Book of Knowledge* has received criticism for its limitations (Abyad, 2012; Gasik, 2011; Koskela & Howell, 2008; Morris, 2013). Since the PMBOK is the leading guide for project management practices in the United States, these implications for practice would be a valuable inclusion.

Recommendations for Further Research

Further research is needed to both validate and expand on the results obtained in this study. First, a repeat of this study could be conducted with a larger sample size. A larger sample size may increase the normality of the data. If not, the sample size should be large enough to conduct the structural equation modeling analysis with weighted least squares, rather than with maximum likelihood estimators. Using a mixed method approach may return interview data that could better explain the survey responses. However, obtaining interview data with a large sample size would be time consuming and costly.

This study could also be repeated controlling for components in the environment in which the project took place. The extent of project management maturity, and the level of regulated methodology, monitoring, and governance used may be a factor. One would expect organizations with well-structured project management capabilities would produce a higher project success level; but if that is true, how much knowledge transfer, and team learning contribute to project success would be worth investigating.

Additionally, the extent of IT infrastructure in the organization could be controlled by adding survey questions regarding the yearly budget for IT expenses. Organizations with a higher budget may have more technical skills and experience available, increasing the effect of knowledge transfer. However, their projects may be larger or more complicated, and therefore more challenged to achieve success. A study controlling for the level of IT in the organization could be beneficial as IT is expanding across all industries.

With the capabilities of communication today, including e-mail and instant messaging, many project teams are not located in the same building, or even in the same state. This study did not address whether the project team was virtual or co-located. A repeat study in this area would determine if there is a greater or lesser effect of knowledge transfer if teams are not co-located.

Moreover, additional research is recommended to understand the relationship between knowledge articulation and cross-project learning. The findings from this study revealed that knowledge articulation was not significantly related to cross-project learning, which was consistent with the research by Newell and Edelman (2008), but surprising. Cross-project learning was found to be significant and positively related to project success, so increasing the knowledge transfer across projects is beneficial. Also, knowledge articulation can be more useful for transferring tacit knowledge, so understanding why it has not been found to benefit cross-project learning is desirable. Additional survey or interview questions that ask what practices are in place to promote cross-project learning may help explain the relationship.

Conclusions

This study was based on the theory of organizational learning and knowledge management. It expanded on the components of dynamic capabilities of knowledge transfer and applied those components to project management in the IT field in an attempt to identify the significance of knowledge transfer and team learning that contribute to project success. There is a gap in the literature on the relationship between organizational learning and project performance (Jugdev & Mathur, 2013; Paramkusham & Gordon, 2013; Reich, 2007). Furthermore, studies have reported that project success rates are low and have consistently been low for many years (Cerpa & Verner, 2009; Hastie & Wojewoda, 2015; Levin, 2010; Lierni & Ribière, 2008). The goal of this study was to ascertain how practitioners may increase the likelihood of project success and to fill the gap in the literature.

A quantitative survey research method was chosen using regression analysis of structural equation modeling. The survey instrument was previously developed by Newell and Edelman (2008), and regarded as one of the advanced instruments in dynamic capabilities (Eriksson, 2013). Qualtrics was used to distribute the survey and a sample size of 128 was obtained. Both SPSS and AMOS software was used to conduct the analysis.

Results showed that the relationship of the variables in the sub-questions was significant and positive with one exception.

A significant relationship was not found between knowledge articulation and cross-project learning. However, this study did not reveal if knowledge articulation did not occur across projects or if it did occur but did not contribute to learning. Further research was recommended to understand this relationship. Results did show a significant and positive relationship between the other variables of knowledge transfer on team learning and on project success. Therefore, it was suggested that organizations develop and support training for individuals to increase individual knowledge. Another suggestion was that time should be allowed for the project team and project stakeholders to develop documentation throughout the project and at project closure to increase the amount of codified knowledge available. A third recommendation for practical project management was to acquire knowledge from past or similar projects to enhance project learning and cross-project learning.

Factors not considered in this study, but may have impacted the results, include the nature of the environment in the organization. Specific factors are the level of project management maturity, the amount of IT infrastructure, and the degree in which the project team members are co-located.

REFERENCES

Abyad, A. (2012). Project mnagement: The challenge, the dilemma. *Middle East Journal of Business*, *7*(1), 18–22. Retrieved from http://www.mejb.com/upgrade_flash/home.htm

Ahern, T., Byrne, P. J., & Leavy, B. (2015). Developing complex-project capability through dynamic organizational learning. *International Journal of Managing Projects in Business*, *8*(4), 732–754. doi: 10.1108/IJMPB-11-2014-0080

Ahern, T., Leavy, B., & Byrne, P. J. (2014). Knowledge formation and learning in the management of projects: A problem solving perspective. *International Journal of Project Management*, *32*(8), 1423–1431. doi: 10.1016/j.ijproman.2014.02.004

Andreassen, T., Lorentzen, B., & Olsson, U. (2006). The impact of non-normality and estimation methods in SEM on satisfaction research in marketing. *Quality and Quantity*, *40*(1), 39–58. doi: 10.1007/s11135-005-4510-y

Blunch, N. (2013). *Introduction to structural equation modeling using IBM SPSS statistics and AMOS* (2nd ed.). London, England: SAGE Publications, Ltd.

Brady, T., & Davies, A. (2014). Managing structural and dynamic complexity: A tale of two projects. *Project Management Journal*, *45*(4), 21–38. doi: 10.1002/pmj.21434

Bresnen, M. (2016). Institutional development, divergence and change in the discipline of project management. *International Journal of Project Management*, *34*(2), 328–338. doi: 10.1016/j.ijproman.2015.03.001

Breznik, L., & Lahovnik, M. (2014). Renewing the resource base in line with the dynamic capabilities view: A key to sustained competitive advantage in the IT industry. *Journal for East European Management Studies, 19*(4), 453–485. doi: 10.1688/JEEMS-2014-04-Breznik

Byrne, B. M. (2010). *Structural eauation modeling with AMOS: Basic concepts, applications, and programming* (2nd ed.). New York, NY: Routledge Taylor & Francis Group.

Cacciatori, E., Tamoschus, D., & Grabher, G. (2011). Knowledge transfer across projects: Codification in creative, high-tech and engineering industries. *Management Learning, 43*(3), 309–331. doi: 10.1177/1350507611426240

Caldwell, R. (2012). Systems thinking, organizational change and agency: A practice theory critique of Senge's learning organization. *Journal of Change Management, 12*(2), 145–164. doi: 10.1080/14697017.2011.647923

Cangelosi, V. E., & Dill, W. R. (1965). Organizational learning: Observations toward a theory. *Administrative Science Quarterly, 10*(2), 175. doi: 10.2307/2391412

Cerpa, N., & Verner, J. M. (2009). Why did your project fail? *Communications of the ACM, 52*(12), 130. doi: 10.1145/1610252.1610286

Chronéer, D., & Backlund, F. (2015). A holistic view on learning in project-based organizations. *Project Management Journal, 46*(3), 61–74. doi: 10.1002/pmj.21503

Cooper, D. R., & Schnider, P. S. (2011). *Business research methods* (11th ed.). New York, NY: McGraw-Hill Irwin.

Crawford, J., Leonard, L. N. K., & Jones, K. (2011). The human resource's influence in shaping IT competence. *Industrial Management & Data Systems*, *111*(2), 164–183. doi: 10.1108/02635571111115128

Crawford, L., & Pollack, J. (2007). How generic are project management knowledge and practice? *Project Management Journal*, *38*(1), 87–97. Retrieved from https://www.pmi.org/learning/publications/project-management-journal

Creswell, J. W. (2009). *Research design: Qualitative, quantitative and mixed method approaches*. Thousand Oaks, CA: SAGE Publications Ltd.

Crossan, M. M., Lane, H. W., & White, R. E. (1999). An organizational learning framework: From intuition to institution. *Academy of Management Review*, *24*(3), 522–537. doi: 10.2307/259140

Crossan, M. M., Maurer, C. C., & White, R. E. (2011). Reflections on the 2009 AMR decade award: Do we have a theory of organizational learning? *Academy of Management Review*, *36*(3), 446–460. doi: 10.5465/AMR.2011.61031806

Danneels, E. (2008). Organizational antecedents of second-order competences. *Strategic Management Journal*, *29*(5), 519–543. doi:org/10.1002/smj.684

Davenport, T. H., De Long, D. W., & Beers, M. C. (1998). Successful knowledge management projects. *Sloan Management Review. Winter98*, *39*(2). doi: 10.1016/j.ygeno.2009.01.004

Davies, A., & Brady, T. (2016). Explicating the dynamics of project capabilities. *International Journal of Project Management*, *34*(2), 314–327. doi: 10.1016/j.ijproman.2015.04.006

Davies, A., Dodgson, M., & Gann, D. (2016). Dynamic capabilities in complex projects: The case of London Heathrow terminal 5. *Project Management Journal*, *47*(2), 26–46. doi: 10.1002/pmj.21574

Davison, R. M., Ou, C. X. J., & Martinsons, M. G. (2013). Information technology to support informal knowledge sharing. *Information Systems Journal, 23*(1), 89–109. doi: 10.1111/j.1365-2575.2012.00400.x

Di Stefano, G., Peteraf, M., & Verona, G. (2014). The organizational drivetrain: A road to integration of dynamic capabilities research. *Academy of Management Perspectives, 28*(4), 307–327. doi: 10.5465/amp.2013.0100

Dillman, D. A., Smyth, J. D., & Christian, L. M. (2009). *Internet, mail, and mixed-mode surveys: The tailored design method. Internet Mail and MixedMode Surveys The tailored design method* (3rd ed.). Hoboken, NJ: John Wiley & Sons, Inc. doi: 10.2307/41061275

Dinur, A. (2011). Tacit knowledge taxonomy and transfer: Case-based research. *Journal of Behavioral & Applied Management, 12*(3), 246–281. Retrieved from https://jbam.scholasticahq.com/

Drouin, N., & Jugdev, K. (2013). Standing on the shoulders of strategic management giants to advance organizational project management. *International Journal of Managing Projects in Business, 7*(1), 61–77. doi: 10.1108/IJMPB-04-2013-0021

Easterby-Smith, M., Crossan, M., & Nicolini, D. (2000). Organizational learning: Debates past, present and future. *Journal of Management Studies, 37*(6), 783–796. doi: 10.1111/1467-6486.00203

Edward, F. (2003). The belmont ethos: The meaning of the belmont principles for human subject protections. *Journal of Research Administration, 34*(2), 19–24. Retrieved from http://srainternational.org/publications/journal/volume-xlvii-number-2

Eisenhardt, K. M., & Martin, J.A. (2000). Dynamic capabilities: What are they? *Strategic Management Journal, 21*, 1105–1121. doi: 10.1002/1097-0266(200010/11)21:10/11<1105::AID-SMJ133>3.0.CO;2-E

Eriksson, T. (2013). Methodological issues in dynamic capabilities research – a critical review. *Baltic Journal of Management*, *8*(3), 306–327. doi: 10.1108/BJOM-Jul-2011-0072

Faul, F., Erdfelder, E., Buchner, A., & Lang, A.-G. (2009). Statistical power analyses using G*Power 3.1: tests for correlation and regression analyses. *Behavior Research Methods*, *41*(4), 1149–60. doi: 10.3758/BRM.41.4.1149

Fenwick, T. (2008). Understanding relations of individual--collective learning in work: A review of research. *Management Learning*, *39*(3), 227–243. doi: 10.1177/1350507608090875

Field, A. (2009). *Discovering Statistics using SPSS* (3rd ed.). Thousand Oaks, CA: SAGE Publications Ltd.

Gasik, S. (2011). A model of project knowledge management. *Project Management Journal*, *42*(3), 23–44. doi: 10.1002/pmj

Gemino, A., Reich, B. H., & Sauer, C. (2008). A temporal model of information technology project performance. *Journal of Management Information Systems*, *24*(3), 9–44. doi: 10.2753/MIS0742-1222240301

Gharaibeh, H. (2011). Improving project team learning in major projects – a case study comparison. *E-Journal of Organizational Learning and Leadership*, *9*(2), 1–13. Retrieved from http://www.leadingtoday.org/weleadinlearning/

Gharaibeh, H. (2012). Project team learning, the mystery revealed. *Journal of Organizational Learning and Leadership*, *10*(2), 1–16. Retrieved from http://www.leadingtoday.org/weleadinlearning/

Goffin, K., & Koners, U. (2011). Tacit knowledge, lessons learnt, and new product development. *The Journal of Product Innovation Management*, *28*, 300–318. doi: 10.1111/j.1540-5885.2010.00798.x

Gonzalez, R. V. D., & Martins, M. F. (2014). Knowledge management: An analysis from the organizational development. *Journal of Technology Management & Innovation, 9*(1), 131–147. doi: 10.4067/S0718-27242014000100011

Gourlay, S. (2006). Conceptualizing knowledge creation: A critique of Nonaka's theory. *Journal of Management Studies, 43*(7), 1415–1436. doi: 10.1111/j.1467-6486.2006.00637.x

Hadaya, P., Cassivi, L., & Chalabi, C. (2012). IT project management resources and capabilities: a Delphi study. *International Journal of Managing Projects in Business, 5*(2), 216–229. doi: 10.1108/17538371211214914

Haldeman, J. (2011). The learning organization: From dysfunction to grace. *Journal of Management & Marketing Research, 9,* 1–9. Retrieved from http://www.aabri.com/jmmr.html

Hall, M., Kutsch, E., & Partington, D. (2012). Removing the cultural and managerial barriers in project-to-project learning: A case from the UK public sector. *Public Administration, 90*(3), 664–684. doi: 10.1111/j.1467-9299.2011.01980.x

Hartmann, A., & Dorée, A. (2015). Learning between projects: More than sending messages in bottles. *International Journal of Project Management, 33*(2), 341–351. doi: 10.1016/j.ijproman.2014.07.006

Hastie, S., & Wojewoda, S. (2015). The Standish Group 2015 chaos report - Q&A with Jennifer Lynch. Retrieved from http://www.infoq.com/articles/standish-chaos-2015

Helfat, C., Finkelstein, S., Michell, W., Peteral, M. A., Singh, H., Teece, D. J., & Winter, S. G. (2007). *Dynamic capabilities: Understanding strategic change in organizations.* Oxford, U.K.: Blackwell.

Hermano, V., & Martin-Cruz, N. (2014). The role of top management involvement in firms performing projects: A dynamic capabilities approach. *Journal of Business Research, 69*(9), 3447–3458. doi: 10.1016/j.jbusres.2016.01.041

Holsapple, C. W. (2003). Knowledge and its attributes. In *Handbook on knowledge management* (pp. 165–188). Berlin, Germany: Springer-Verlag.

Holzmann, V. (2013). A meta-analysis of brokering knowledge in project management. *International Journal of Project Management, 31*(1), 2–13. doi: 10.1016/j.ijproman.2012.05.002

Islam, M. Z., Low, K. C. P., & Rahman, M. H. (2012). Towards understanding knowledge transfer: In search of a theretical construct. *Franklin Business & Law Journal, 2012*(1), 39–61. Retrieved from http://www.franklinpublishing.net/businesslaw.html

Jackson, D. L. (2003). Revisiting sample size and number of parameter estimates: Some support for the N:q hypothesis. *Structural Equation Modeling, 10*(1), 128–141. doi: 10.1207/S15328007SEM1001_6

Jackson, P., & Klobas, J. (2008). Building knowledge in projects: A practical application of social constructivism to information systems development. *International Journal of Project Management, 26*(4), 329–337. doi: 10.1016/j.ijproman.2007.05.011

Jetu, F. T., & Riedl, R. (2012). Determinants of information systems and information technology project team success: A literature review and a conceptual model. *Communications of the Association for Information Systems, 30*(1), 455–482. Retrieved from http://aisel.aisnet.org/cais/vol30/iss1/27

Jewels, T., & Ford, M. (2006). Factors influencing knowledge in information technology projects. *E-Service Journal, 5*(1), 99–117. doi: 10.2979/ESJ.2006.5.1.99

Jugdev, K. (2012). Learning from lessons learned: Project management research program. *American Journal of Economics and Business Administration, 4*(1), 13–22. doi: 10.3844/ajebasp.2012.13.22

Jugdev, K., & Mathur, G. (2013). Bridging situated learning theory to the resource-based view of project management. *International Journal of Managing Projects in Business, 6*(4), 633–653. doi: 10.1108/IJMPB-04-2012-0012

Karimi, J., & Walter, Z. (2015). The role of dynamic capabilities in responding to digital disruption: A factor-based study of the newspaper industry. *Journal of Management Information Systems, 32*(1), 39–81. doi: 10.1080/07421222.2015.1029380

Kline, R. B. (2011). *Principles and practice of structural equation modeling* (3rd ed.). New York, NY: The Guilford Press.

Koskela, L., & Howell, G. (2008). The underlying theory of project management is obsolete. *IEEE Engineering Management Review, 36*(2), 22–34. doi: 10.1109/EMR.2008.4534317

Krzakiewicz, K. (2013). Dynamic capabilities and knowledge management. *Management, 17*(2), 1–15. doi: 10.2478/manment-2013-0051

Landaeta, R. E. (2008). Evaluating benefits and challenges of knowledge transfer across projects. *Engineering Management Journal, 20*(1), 29–38. doi: 10.1080/10429247.2008.11431753

Lave, J., & Wenger, E. (1991). *Situated learning: Legitimate peripheral participation.* New York, NY: Cambridge University Press.

Levin, G. (2010, October). Knowledge management success equals project management success. In *PMI Global Congress* (Vol. 11).

Leybourne, S., & Kennedy, M. (2015). A process-based classification of knowledge maps and application examples. *Knowledge and Process Management, 22*(1), 1–10. doi: 10.1002/kpm.1457

Lichtenthaler, U. (2009). Absorptive capacity, environmental turbulence, and the complementarity of organizational learning processes. *Academy of Management Journal, 52*(4), 822–846. doi: 10.5465/AMJ.2009.43670902

Lierni, P. C., & Ribière, V. M. (2008). The relationship between improving the management of projects and the use of KM. *Vine, 38*(1), 133–146. doi: 10.1108/03055720810870941

Lim, J.H., Stratopoulos, T. C., & Wirjanto, T. S. (2012). Path dependence of dynamic information technology capability: An empirical investigation. *Journal of Management Information Systems, 28*(3), 45–84. doi: 10.2753/MIS0742-1222280302

Lin, Y., & Wu, L. Y. (2014). Exploring the role of dynamic capabilities in firm performance under the resource-based view framework. *Journal of Business Research, 67*(3), 407–413. doi: 10.1016/j.jbusres.2012.12.019

Liu, D., Ray, G., & Whinston, A. B. (2010). The interaction between knowledge codification and knowledge-sharing networks. *Information Systems Research, 21*(4), 892–906. doi: 10.1287/isre.1080.0217

Luhman, J. T., & Cunlifffe, A. L. (2013). Organizational learning and knowledge management. In *SAGE Key Concepts series: Key Concepts in Organization Theory* (pp. 124–133). London, England: SAGE Publications Ltd. doi: 10.1111/j.1468-2370.2009.00257.x

Mahaney, R. C., & Lederer, A. L. (2011). An agency theory explanation of project success. *Journal of Computer Information Systems, 51*(4), 102–113. Retrieved from http://www.iacis.org/

Markus, M. L. (2001). Toward a theory of knowledge reuse: Types of knowledge reuse situations and factors in reuse success. *Journal of Management Information Systems, 18*(1), 57–93. Retrieved from http://www.jmis-web.org/issues

McKay, D. S. (2012). *The Interactions Among Information Technology Organizational Learning, Project Learnin, and Project Success* (Doctoral dissertation).). Retrieved from ProQuest Central. ((Order No. 3517557.

Morris, P. (2013). Reconstructing project management reprised: A knowledge perspective. *Project Management Journal, 44*(5), 6–23. doi: 10.1002/pmj.21369

Mueller, J. (2015). Formal and informal practices of knowledge sharing between project teams and enacted cultural characteristics. *Project Management Journal, 46*(1), 53–68. doi: 10.1002/pmj.21471

Naldi, L., Wikström, P., & Von Rimscha, M. B. (2014). Dynamic capabilities and performance. *International Studies of Management and Organization, 44*(4), 63–82. doi: 10.2753/IMO0020-8825440404

Newell, S., Bresnen, M., Edelman, L., Scarbrough, H., & Swan, J. (2006). Sharing knowledge across projects: Limits to ICT-led project review practices. *Management Learning, 37*(2), 167–185. doi: 10.1177/1350507606063441

Newell, S., & Edelman, L. F. (2008). Developing a dynamic project learning and cross-project learning capability: Synthesizing two perspectives. *Information Systems Journal, 18*(6), 567–591. doi: 10.1111/j.1365-2575.2007.00242.x

Nonaka, I. (1991). The knowledge-creating company. *Harvard Business Review, 69*(6), 96–104. doi: 10.1016/0024-6301(96)81509-3

Nonaka, I. (1994). A dynamic theory of organizational knowledge creation. *Organization Science, 5*(1), 14–37. doi: 10.1057/9781137024961_5

Nonaka, I., & Toyama, R. (2003). The knowledge-creating theory revisited: knowledge creation as a synthesizing process. *Knowledge Management Research & Practice, 1*(1), 2–10. doi: 10.1057/palgrave.kmrp.8500001

Nonaka, I., & Von Krogh, G. (2009). Tacit knowledge and knowledge conversion: Controversy and advancement in organizational knowledge creation theory. *Organization Science, 20*(3), 635–652. doi: 10.1287/orsc.1080.0412

Paramkusham, R. B., & Gordon, J. (2013). Inhibiting factors for knowledge transfer in information technology projects. *Journal of Global Business & Technology, 9*(2), 26–37. Retrieved from http://gbata.org/journal-of-global-business-and-technology-jgbat/publications/

Penrose, E. T. (1959). The theory of the growth of the firm. *New York: Sharpe.*

Peteraf, M., Di Stefano, G., & Verona, G. (2013). The elephant in the room of dynamic capabilities: Bringing two diverging conversations together. *Strategic Management Journal, 34*(12), 1389–1410. doi: 10.1002/smj.2078

Pinto, J. K., & Slevin, D. P. (1988a). Critical success factors across the project life cycle. *Project Management Journal, 19*(3), 67–75. Retrieved from pmi.org.

Pinto, J. K., & Slevin, D. P. (1988b). Project success - definitions and measurement techniques. *Project Management Journal, 19*(1), 67–72. Retrieved from pmi.org.

Polanyi, M. (1966). *The Tacit Dimension.* Chicago and London: The University of Chicago Press.

Project Management Institute. (2008). *A guide to the project management body of knowledge (PMBOK Guide)* (4th ed.). Newtown Square, PA: Project Management Institute, Inc.

Project Management Institute. (2013). Project management between 2010 +2020. *Project Management Talent Gap Report.* Retrieved from www.pmi.org/-/media/pmi/documents/.../project-management-skills-gap-report.pdf

Project Management Institute. (2015). *Earning power: Project management salary survey* (9th ed.). Newtown Square, PA: Project Management Institute.

Project Management Institute. (2016). The high cost of low performance: How will you improve business results? *PMI's Pulse of the Profesion.* Retrieved from www.pmi.org/-/media/pmi/documents/public/.../pulse-of-the-profession-2016.pdf

Reich, B. H., Gemino, A., & Sauer, C. (2008). Modeling the knowledge perspective of IT projects. *Project Management Journal, 39*(S1), S4–S14. doi: 10.1002/pmj.20056

Reich, B. H., Gemino, A., & Sauer, C. (2012). Knowledge management and project-based knowledge in it projects: A model and preliminary empirical results. *International Journal of Project Management, 30*(6). doi: 10.1016/j.ijproman.2011.12.003

Reich, B. H., & Wee, S. Y. (2006). Searching for knowledge in the PMBOK guide. *Project Management Journal, 37*(2000), 11–26. Retrieved from https://www.pmi.org/learning/publications/project-management-journal

Reich, B. H. (2007). Managing knowledge and learning in IT projects: A conceptual framework and guidelines for practice. *Project Management Journal, 38*(2), 5–17. Retrieved from https://www.pmi.org/learning/publications/project-management-journal

Rovai, A. P., Baker, J., & Ponton, M. K. (2014). *Social science research design and statistics: A practioner's guide to research methods and IBM SPSS analysis* (1st ed.). Chesapeake, VA: Watertree Press LLC.

Rungi, M. (2014). The impact of capabilities on performance. *Industrial Management & Data Systems, 114*(2), 241–257. doi: 10.1108/IMDS-04-2013-0202

Salunke, S., Weerawardena, J., & McColl-Kennedy, J. R. (2011). Towards a model of dynamic capabilities in innovation-based competitive strategy: Insights from project-oriented service firms. *Industrial Marketing Management, 40*(8), 1251–1263. doi: 10.1016/j.indmarman.2011.10.009

Scarso, E., & Bolisani, E. (2011). Managing professions for knowledge management. *International Journal of Knowledge Management, 7*(3), 61–75. doi: 10.4018/jkm.2011070105

Scurtu, L. E., & Neamtu, D. M. (2015). The need of using knowledge management strategy in modern business organizations. *USV Annals of Economics and Public Administration, 15*(2), 157–167. Retrieved from http://www.seap.usv.ro/annals/ojs/index.php/annals

Senge, P. M. (1994). *The fifth discipline: The art & practice of the learning organization.* New York, NY: Currency Doubleday.

Singh, A., & Soltani, E. (2010). Knowledge management practices in Indian information technology companies. *Total Quality Management & Business Excellence, 21*(2), 145–157. doi: 10.1080/14783360903549832

Slepian, J. L. (2013). Cross-functional teams and organizational learning: A model and cases from telecommunications operating companies. *International Journal of Innovation and Technology Management, 10*(01), 1350005. doi: 10.1142/S0219877013500053

Sproull, N. L. (2003). *Handbook of research methods: A guide for practitioners and students in the social sciences* (2nd ed.). Berlin, Germany: Springer-Verlag.

Standish Group. (2013). CHAOS MANIFESTO 2013: Think big, act small. *The Standish Group International.* Retrieved from http://www.versionone.com/assets/img/files/CHAOSManifesto2013.pdf

Steenkamp, A.L., & McCord, S. A. (2003). Approach to teaching research methodology for information technology. *Journal of Information Systems, 18*(2), 255–267. doi: 1324109781.

Stewart, W. H., May, R. C., & Ledgerwood, D. E. (2015). Do you know what I know? Intent to share knowledge in the US and Ukraine. *Management International Review, 55*(6), 737–773. doi: 10.1007/s11575-015-0252-9

Steyn, C., & Kahn, M. (2008). Towards the development of a knowledge management practices survey for application in knowledge intensive organisations. *South African Journal of Business Management, 39*(1), 45–53. Retrieved from http://journals.co.za/content/journal/busman

Swan, J., Scarbrough, H., & Newell, S. (2010). Why don't (or do) organizations learn from projects? *Management Learning, 41*(3), 325–344. doi: 10.1177/1350507609357003

Teece, D. J. (2007). Explicating dynamic capabilities: the nature and microfoundations of (sustainable) enterprise performance. *Strategic Management Journal, 28*(13), 1319–1350. doi: 10.1002/smj.640

Teece, D. J. (2014). The foundations of enterprise performance: Dynamic and ordinary capabilities in an (economic) theory of firms. *Academy of Management Perspectives, 28*(4), 328–352. doi: 10.5465/amp.2013.0116

Teece, D. J., Pisano, G., & Shuen, A. M. Y. (1997). Dynamic capabilities and strategic management. *Strategic Management Journal, 18*(7), 509–533. doi: 10.1002/(sici)1097-0266(199708)18:7<509::aid-smj882>3.0.co;2-z

Timbrell, G., Delaney, P., Chan, T., Yue, A., & Gable, G. (2005). A structurationist review of knowledge menagement theories. In *Twenty-Sixth International Conference on Information Systems* (pp. 247–259). Atlanta: Association for Information Systems.

Trochim, W., & Donnelly, J. (2006). *The research methods knowledge base* (3rd ed.). Mason, OH: Cengage Learning.

Tzortzaki, A. M., & Mihiotis, A. (2014). A review of knowledge management theory and future directions. *Knowledge and Process Management, 21*(1), 29–41. doi: 10.1002/kpm.1429

Vogt, W. (2007). *Quantitative research methods for professionals.* Boston, MA: Pearson Education.

von Krogh, G. (2002). The communal resource and information systems. *The Journal of Strategic Information Systems, 11*(2), 85–107. doi: 10.1016/S0963-8687(02)00006-9

von Krogh, G., Nonaka, I., & Rechsteiner, L. (2012). Leadership in organizational knowledge creation: A review and framework. *Journal of Management Studies, 49*(1), 240–277. doi: 10.1111/j.1467-6486.2010.00978.x

Wiewiora, A., Murphy, G., Trigunarsyah, B., & Brown, K. (2014). Interactions between organizational culture, trustworthiness, and mechanisms for inter-project knowledge sharing. *Project Management Journal, 45*(2), 48–65. doi: 10.1002/pmj.21407

Wu, L. Y. (2010). Applicability of the resource-based and dynamic-capability views under environmental volatility. *Journal of Business Research, 63*(1), 27–31. doi: 10.1016/j.jbusres.2009.01.007

Zhao, D., Zuo, M., & Deng, X. (2011). Examining the influencing factors of cross-project knowledge transfer: An empirical study of IT service firms. *International Conference on Information Systems*, 1–12. doi: 10.1016/j.ijproman.2014.05.003

Zheng, S., Zhang, W., & Du, J. (2011). Knowledge-based dynamic capabilities and innovation in networked environments. *Journal of Knowledge Management, 15*(6), 1035–1051. doi: 10.1108/13673271111179352

Zollo, M., & Winter, S. G. (2002). Deliberate learning and the evolution of dynamic capabilities. *Organization Science, 13*(3), 339–351. doi: 10.1287/orsc.13.3.339.2780

STATEMENT OF ORIGINAL WORK

Academic Honesty Policy

Capella University's Academic Honesty Policy (3.01.01) holds learners accountable for the integrity of work they submit, which includes but is not limited to discussion postings, assignments, comprehensive exams, and the dissertation or capstone project.

Established in the Policy are the expectations for original work, rationale for the policy, definition of terms that pertain to academic honesty and original work, and disciplinary consequences of academic dishonesty. Also stated in the Policy is the expectation that learners will follow APA rules for citing another person's ideas or works.

The following standards for original work and definition of plagiarism are discussed in the Policy:

> Learners are expected to be the sole authors of their work and to acknowledge the authorship of others' work through proper citation and reference. Use of another person's ideas, including another learner's, without proper reference or citation constitutes

> plagiarism and academic dishonesty and is prohibited conduct. (p. 1)
>
> Plagiarism is one example of academic dishonesty. Plagiarism is presenting someone else's ideas or work as your own. Plagiarism also includes copying verbatim or rephrasing ideas without properly acknowledging the source by author, date, and publication medium. (p. 2)

Capella University's Research Misconduct Policy (3.03.06) holds learners accountable for research integrity. What constitutes research misconduct is discussed in the Policy:

> Research misconduct includes but is not limited to falsification, fabrication, plagiarism, misappropriation, or other practices that seriously deviate from those that are commonly accepted within the academic community for proposing, conducting, or reviewing research, or in reporting research results. (p. 1)

Learners failing to abide by these policies are subject to consequences, including but not limited to dismissal or revocation of the degree.

Statement of Original Work and Signature

I have read, understood, and abided by Capella University's Academic Honesty Policy (3.01.01) and Research Misconduct Policy (3.03.06), including Policy Statements, Rationale, and Definitions.

I attest that this dissertation or capstone project is my own work. Where I have used the ideas or words of others, I have paraphrased, summarized, or used direct quotes following the guidelines set forth in the APA Publication Manual.

Learner name and date: Dixie O'Connell Overton — February 8, 2017

INDEX

Agency Theory 7, 78

Behavior Theory 7, 77

Behavior 9, 11, 22, 41, 48, 51-53, 58-59, 61-62, 64-65, 72, 108

Belmont Report 108-110

Bias 27, 28, 116, 144

Capella University 28, 100-102, 109, 178-180

Cronbach's Alpha 104-105, 108, 117-118, 149-150

Cross-Project Learning 8, 17-19, 31, 52, 66-70, 83, 87-88, 93, 102, 104-106, 108, 116-118, 123-126, 128-129, 131-140, 142-143, 145-147, 149-154, 156-157, 159, 161

Culture 7, 28, 34, 38, 50, 63-64, 82

Data Analysis 27-29, 99, 103, 111, 117

Data Collection 28-29, 89, 93, 101, 103, 109, 112

Demographic 19-21, 24-25, 95-97, 102, 110, 112

Dynamic Capabilities 10, 14, 16, 19, 26, 29-32, 36, 42-47, 52, 54-59, 72, 74-78, 80-81, 83, 85-86, 89, 142-143, 148-149, 152, 160

Experience Accumulation 20, 58, 85, 150

Explicit Knowledge 4, 20, 35-39, 53, 60, 62, 85, 92

Individual Knowledge 2, 8, 10, 14-15, 17-18, 20, 24, 31, 58-60, 72, 83, 85-87, 91, 105-106, 108, 112, 116-118, 123-124, 126-136, 138-140, 143, 145-147, 149-153, 155-156, 161

Information Technology (IT) 1, 11-12, 14, 17, 21-22, 33, 47-48, 54, 107, 124, 140, 143, 146

Knowledge Articulation 10, 14-15, 17-18, 22, 24, 31, 54, 56, 58-60, 62-63, 72, 76, 83, 85-87, 91-92, 105-106, 108, 112, 116-118, 123-125, 129-136, 138-140, 143, 145-147, 149-151, 153-156, 159, 161

Knowledge-Based View 32, 35, 64, 74, 77, 80, 83, 152

Knowledge Codification 10, 13-15, 17-19, 23-24, 31, 54, 56, 58-63, 68, 72, 83, 85-88, 91-92, 105-106, 108, 112, 116-118, 123-125, 132-136, 138-140, 143, 145-147, 150-152, 156-157

Knowledge Creation 23, 33, 37-38, 40, 42, 56, 65, 151

Knowledge Creation Theory 8, 10, 30-31, 37, 39, 142

Knowledge Documentation 13, 23, 76

Knowledge Management 1-6, 8, 13-17, 23, 30-32, 34-37, 41, 43, 46-47, 49-51, 55-56, 61, 63, 82-86, 142-143, 157, 160

Knowledge Sharing 2, 4, 9, 23, 33, 35, 38, 48, 50-51, 53-55, 57, 62-65, 67, 69, 77, 90, 92

Knowledge Transfer 2-3, 7, 9-17, 23-24, 30, 36, 47, 51, 53, 59, 62, 68-69, 78, 82, 85-86, 90-91, 93, 105, 112, 124, 135, 137-138, 140, 142-144, 146-147, 149, 155-156, 158-161

Learning Organization 7, 33, 40, 65

Lessons Learned 13, 20, 24, 61-62, 68, 76, 92, 151

Likert-Type 11, 103, 106, 116

Network Ties 62-65, 156

Organizational Learning 3, 5-10, 13-14, 17, 23-24, 30-34, 36, 40, 43, 47-49, 55, 58, 64-65, 67, 76-77, 83, 92, 142, 148, 155, 160

Organizational Learning Theory 7-8, 32-33, 82, 142

Project Management 1-6, 8, 10-11, 13-16, 25, 27, 30-32, 41, 47-55, 59, 66, 70, 73-74, 76, 78-80, 82-86, 89, 95-97, 141-144, 148-149, 155, 157-158, 160-161

Project Management Book of Knowledge 4, 24, 47-48, 52, 61, 85, 157

Project Manager 3, 5, 11, 16-17, 25, 73, 113, 141, 144, 150, 155-157

Project Success 1-3, 6-9, 11, 13-17, 19, 25, 29, 47, 53, 64, 68, 70-73, 75-76, 78-83, 85-86, 88, 90-93, 102, 105-106, 108, 112, 116-118, 123-124, 126, 135-140, 142-147, 149-154, 156-161

Project Team 4, 6, 13, 19, 23, 25, 28, 32, 38, 49-50, 54, 59-60, 64-71, 73, 75, 78, 80-81, 92-93, 97, 113-115, 145-147, 150-153, 155-157, 159, 161

Project Team Learning 2, 4, 8, 14, 31, 52, 64-65, 67-68, 70, 102, 142, 154

Qualtrics 21, 26-28, 89, 97, 100-103, 110, 160

Quantitative 11, 26-27, 70, 74-75, 82, 112, 155, 160

Regression 11, 15, 17, 27, 68, 85, 86, 99, 104, 108, 119-121, 126-138, 160

Reliability 108, 111-112, 117-118, 145

Resource-Based View (RBV) 32, 34-35, 42, 45, 47, 73-74, 76, 83

Resource-Based Theory 35, 42, 44

SECI-Ba 36-38, 40

Sharing Potential 63

Situated Learning 30, 40-41, 51, 61, 83

Social 34, 36-37, 40-42, 48-49, 54, 63, 65-66, 69, 80, 83

Social Research 11

Structural Equation Modeling (SEM) 27, 74-75, 91, 93, 99, 104, 108, 119, 143-144, 148, 158, 160

Structuration Theory 41-42

Survey Design 89

Survey Instrument 27-29, 74-75, 79, 83-84, 89-90, 104-105, 109, 111, 118, 142, 145, 149, 160

Tacit Knowledge 26, 35-40, 52-53, 58-60, 62, 69, 71, 78, 85, 91, 156, 159

Trust 34, 64, 80-81, 102, 152-153

Validity 105-107, 111-112, 118, 145, 149

Variables 10-12, 17, 19-27, 30-31, 47, 51, 58, 64, 81, 83-84, 86, 90-93, 9899, 101, 103-106, 108, 116-124, 127, 130, 133, 138, 143-145, 149-150, 160-161

www.ingramcontent.com/pod-product-compliance
Lightning Source LLC
Chambersburg PA
CBHW030938180526
45163CB00002B/616